# DENDROSCAN: A TREE-RING WIDTH AND DENSITY MEASUREMENT SYSTEM

*T.M.L. Varem-Sanders and I.D. Campbell*

SPECIAL REPORT 10

Canadian Forest Service
Northern Forestry Centre
1996

Catalogue No. Fo29-34/10-1996E
ISBN 0-660-16659-3
ISSN 1188-7419

Distributed by:

UBC Press
University of British Columbia
6344 Memorial Road
Vancouver, British Columbia V6T 1Z2
Phone (604) 822-5959
Fax 1-800-668-0821

PRINTED IN CANADA
IMPRIMÉ AU CANADA

CANADIAN CATALOGUING IN PUBLICATION DATA

Varem-Sanders, T.M.L. (Thierry Marie Louis), 1954-

DendroScan: a tree-ring width and density measurement system

(Special report / Canadian Forest Service, Northern Forestry Centre ; 10)
Includes bibliographical references and index.
ISBN 0-660-16659-3
CCG cat. no. Fo29-34/10-1996E

1. DendroScan (Computer software) 2. Dendrochronology — Computer software. 3. X-ray densitometry in dendrochronology — Computer software. I. Campbell, Ian D., 1962- II. Northern Forestry Centre (Canada) III. Title. IV. Series: Special report (Northern Forestry Centre (Canada)) ; 10.

QK477.2.A6V37 1996 582.16 C96-901083-4

**Varem-Sanders, T.M.L.; Campbell, I.D. 1996. DendroScan: a tree-ring width and density measurement system. Nat. Resour. Can., Can. For. Serv., North. For. Cent., Edmonton, Alberta. Spec. Rep. 10.**

# ABSTRACT

DendroScan is a system developed for measuring tree-ring width and density. Tree rings contain a record of diameter growth and carbon storage in the tree trunk. The width of the rings can be measured using a variety of systems, but the density (carbon storage) is most-easily measured by X-ray densitometry. The DendroScan system enables the user to measure both tree-ring width and density, or just ring widths. The system can use prepared wood samples (for ring widths only) or X-ray images. DendroScan is fully described, with discussion of sample preparation, calibration procedures, and data analysis. Information and instructions for the user are included. The DendroScan program is enclosed on a DOS-formatted disk, along with a precalibrated precision wedge for converting X-ray images to density.

# RÉSUMÉ

DendroScan est un système développé pour mesurer les épaisseurs et densités des cernes annuels des arbres. Les cernes enregistrent la croissance et stockage de carbone annuels du tronc de l'arbre. L'épaisseur des cernes peut être mesuré selon une variété de techniques, mais la densité (stockage de carbone) est le plus facilement estimé par radiographie. Le système DendroScan permet de mesurer simultanément les épaisseurs et densités des cernes, ou de mesurer seulement les épaisseurs des cernes. Le système peut s'appliquer soit à des tranches de bois préparées (pour les épaisseurs seulement) ou à des radiographies. Ce manuel décrit en detail le système DendroScan, la préparation des échantillons, la calibration du système, et l'analyse préliminaire des données. Des informations et instructions à l'intention de l'usager sont inclues. Le program DendroScan est inclus sur une disquette format DOS, ainsi qu'un coin de calibration à haute précision pour la conversion des radiographies en valeurs de densité.

# CONTENTS

# FIGURES



# TABLES

# ACKNOWLEDGMENTS

DendroScan has benefited from the insight and suggestions of many people. Tedd Hogg, Rick Hurdle, and Dan MacIsaac contributed to the devlopment of the program and the concept. Glen MacDonald and Kent Holden provided useful critical reviews of various drafts of both the manual and the software. Colin Myrholm, Hanita Brungs, Peter Tollestrup, Andrew Dickinson, Peter Englefield, and Charlotte Varem-Sanders served as beta testers, often enduring maddening system crashes and inexplicable computer behavior. Elaine Schiewe designed and typeset the manuscript and Brenda Laishley coordinated the editing and production of the final product.


## NOTE

*The exclusion of certain manufactured products does not necessarily imply disapproval nor does the mention of other products necessarily imply endorsement by Natural Resources Canada.*

# INTRODUCTION

The usefulness of tree-ring width chronologies is well-known and many researchers around the world are engaged in studies involving ring widths. Ring densities also carry a great deal of information, but fewer researchers measure ring densities, in part due to the cost of the specialized equipment required for measurement.

In 1992, scientists determined that X-ray densitometric data would be beneficial to research work at the Northern Forestry Centre (NoFC) in Edmonton, Alberta. Accordingly, various approaches to obtaining that data were considered. The first test used clinical mammography. These X-ray images had insufficient resolution (0.5 mm) for analysis due to the effect of medical intensifying screens. At about this time, an X-ray inspection system was acquired.

The next hurdle was to develop equipment and techniques for accurately and consistently slicing sufficiently thin wood samples. Sanding the samples introduced sanding dust and grit into the wood pores (a minor difficulty, as most could be removed with further treatment), but also produced samples of uneven thickness. Eventually, a twin-blade modification to a table saw was developed that proved adequate to the task.

The final challenge was analyzing the X-ray images with high resolution and precision. A desktop scanner could digitize the image, and gray values were converted to density using a continuous plastic calibration wedge. A digital-analysis method and computer program were developed to assist in the translation of the image.

This manual presents the details of the system in its final, working configuration, so that others wishing to assemble similar systems need not repeat much of the time-consuming trial-and-error of developing a system from scratch. Also included is the digital analysis program created by the authors (DendroScan) and a pre-calibrated X-ray wedge.

The discussion is separated into four parts: System Overview, which summarizes the basic procedures and associated measurement precision particular to this system; Sample Preparation, which describes how to collect and prepare increment core and disk samples for X-ray or direct scanning; Scanning and Marking Images, which describes the methodology for scanning and preparing images for digital analysis; and DendroScan Users' Manual, which describes the digital analysis program enclosed. Appendixes 1 through 7 provide additional information about DendroScan for the user, including trouble-shooting tips.

# PART 1 SYSTEM OVERVIEW

The system developed at Natural Resources Canada's Northern Forestry Centre (NoFC) has the advantages of being inexpensive and composed mainly of off-the-shelf parts available in many models from different manufacturers. The system has the potential to measure ring density with an accuracy of ±1%, and ring widths within a maximum accuracy of 0.004 mm. The software (DendroScan) can be used to measure ring widths on raw wood samples, or ring widths and densities on X-rayed samples.

## 1.1 Indirect vs. Direct X-ray Reading Systems

There are two general types of X-ray densitometry systems. The first type (direct reading) involves guiding a restricted X-ray beam through the sample to a radiation detector, which is connected to a computer capable of interpreting and recording the X-ray signal. A recently assembled direct-reading system is described by Jozsa and Myronuk (1986). While the direct-reading X-ray densitometry system has the advantage of conceptul and operational simplicity, small space requirements, and few sources of error, it requires a large capital outlay for hardware, a highly stable X-ray source and detector, and primarily custom-built data-logging facilities for translating the output.

The DendroScan system is of the second type (*see* Fig. 1.1.1), which is an indirect reading system. This system involves exposing an X-ray film (negative) that can be scanned directly or used to create a positive print that is then scanned. The gray values of the scanned image correspond to X-ray densities on the film. This indirect or film-stage type system is described in various publications, including Fritts (1976). The capital costs of this system are probably lower because the equipment (X-ray chamber, scanner, darkroom, and computer facilities) is readily available and relatively inexpensive. Operating costs are higher, however, due to film and developer consumption and operator time.

## 1.2 Precision Cutting Samples

X-ray densitometry requires that the samples be sliced to a thickness of 1–2 mm with as little variation as possible both within and between samples. Special saws have been designed for this purpose (Jozsa and Myronuk 1986).

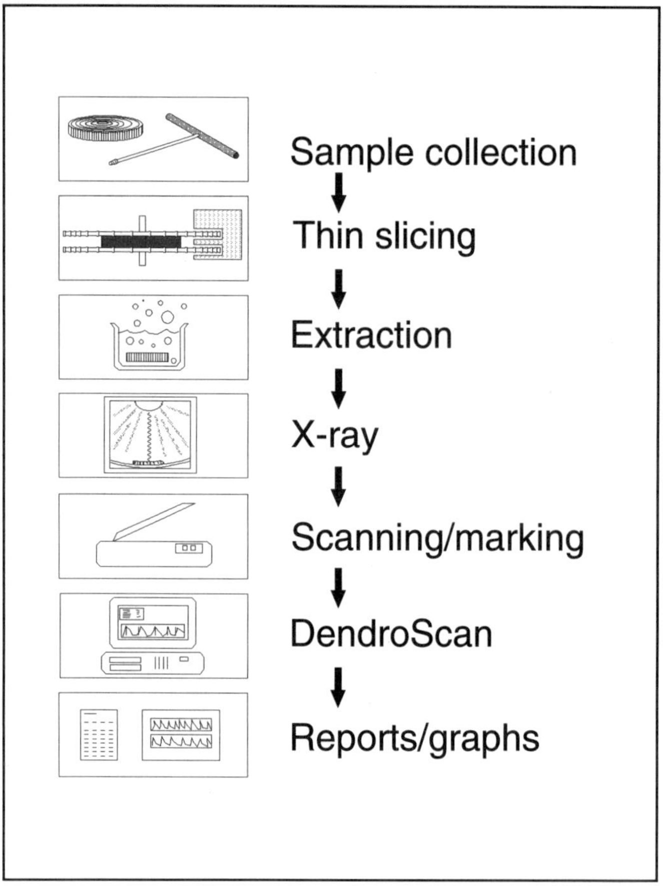


**Figure 1.1.1 DendroScan system flowchart.**

They typically cost approximately $10,000.[1] The main features of these special saws are: high rim-speed to produce a large number of individual tooth impacts per second; small blades to reduce wobble; milling blades (in which the teeth do not protrude to the sides as in most saw blades) to minimize friction; pneumatic power to reduce vibration; twin blades at a variable separation but rigidly parallel; and a clamp/jig for rigidly holding the samples (primarily increment cores). All these features combine to minimize vibration of the saw or sample and allow a very precise cut. The clamp is usually

[1] All prices cited in this report are in Canadian dollars.

designed to hold increment cores securely. Samples taken from disks must first be reduced to fit in the mounting clamp.

The system designed by the authors employs a standard table saw equipped with two carbide-tipped blades reinforced with blade stiffeners and separated by a precision-machined separator disk. The samples are securely held using a key-slot mounted jig. The combination of these modifications minimizes vibration and friction, and results in precise and repeatable sample slices for a small capital cost. In contrast to the clamps on the special saws, the table saw-jig holds block-sized pieces of wood rather than increment cores. As a result, samples are easily taken from cross-sectional disks, but increment cores require preparation prior to cutting. They must be first mounted in a slotted wooden block, and then cut.

## 1.3 Calibration and Digital Analysis

In order to measure wood densities, the X-ray absorption must be measured and calibrated. This is fairly straightforward on direct reading systems as they can be hard-calibrated. Indirect systems require that each individual image be calibrated because it is difficult to adequately control all of the variables affecting imaging, development, and scanning. Some indirect systems use a stepped wedge made of plastic sheets to convert gray scale to density.

DendroScan uses a precisely cast epoxy resin calibration wedge that must be included in each X-ray image. The X-ray negatives are used to produce positive photographs that are then digitally scanned to produce a bit-map computer image. DendroScan automatically performs the calibration of both density and spatial resolution for each image using the known dimensions of the calibration wedge. Before the advent of computer-image analysis, this procedure would have been both tedious and time-consuming. DendroScan reduces this process to an almost transparent procedure.

## 1.4 Analytical Precision

Both density and spatial dimension (i.e., ring widths) are measured by X-ray densitometry. The resolution of the system is limited by the following factors: quality of the sample, its collection and preparation; grain size of the X-ray negative; percentage enlargement and grain size of the positive image; and the digital resolution (and importantly, quality) of the scanner.

In all X-ray densitometry systems, the highest precision is achieved when the tracheid and/or vessel walls are precisely aligned between the X-ray source and detector. Furthermore, twisting of the wood grain can result in variable cell alignment within one sample, resulting in loss of resolution. The resolution can therefore be maximized by collecting samples with care (*see* Part 2 of this manual), taking replicate samples, and slicing the samples as thinly as possible.

The authors use a high-resolution, single-emulsion X-ray film (AGFA D3SC). The grain size of this film is fine enough that details as small as individual cell walls are clearly visible under a microscope. Prints are routinely produced with an enlargement factor of two, and scanned at a resolution of 600 dots per inch (dpi), yielding digital images with a spatial resolution of 0.02 mm. Under special circumstances, prints with an enlargement factor of 10 have been produced, yielding an ultimate spatial resolution of 0.0042 mm. The usual resolution of 0.02 mm was found to be adequate when the smallest rings in a given sample are wider than 0.1 mm. Greater enlargements are only necessary for unusually tight-ringed samples.

The scanned image contains 256 gray scale values, providing a potential density accuracy of ±0.3%. The relationship between gray scale and equivalent wood density, however, is nonlinear (*see* Part 3 for details). The actual error varies from 0.05% to ±2.00%, depending on the density range being measured. DendroScan compensates for this to some degree by averaging pixels parallel to the rings in the sample. The actual standard error has been estimated at 0.0093 $g/cm^3$, which equates to an approximate relative error of ±1.0% across most of the density range.

Most scanners introduce a slight smoothing in the density, mainly due to internal reflections and limitations in the detector optics. The gray value sensed at one pixel location might be slightly influenced by the value of its neighboring pixels. This smoothing is difficult to evaluate, but appears to have the effect of a heavily center-weighted moving average.

Small systematic procedural biases in density estimates can occur due to the complexity of the process. Most of these biases, however, are consistent throughout an image. The result is that, although the random error in density might vary only 1% within a sample, the density of the entire sample might be biased by as much as 5% compared to gravimetric density. This is not critical in many studies, as the relative density is unaffected. For biomass studies, though, it might be necessary to measure the bulk density of the samples gravimetrically, and apply a correction factor to each sample to cancel the bias.

## 1.5 Advantages of Digital Image Processing—DendroScan

DendroScan is easy-to-use, menu-driven software that can accurately measure ring widths (X-rayed or raw wood samples) and densities (X-rayed samples only). It is not tied to a specific hardware because it operates on a standard digital-image format (Tagged Image File Format [TIFF or TIF]). It can therefore analyze samples digitized by different scanners at any resolution. The user can define the analytical path along the samples simply by drawing digital lines perpendicular to the ring boundaries, providing for accurate radial-width measurements even when the ring angle changes. DendroScan translates the image along these paths into a linear density signal, in which it identifies ring boundaries and measures ring widths automatically.

The ring identification algorithm in DendroScan is highly accurate with conifers, but less-reliable with some diffuse porous woods such as *Populus*. The variable cell sizes throughout diffuse porous rings make ring boundary identification more difficult using density alone. Typically, DendroScan will correctly identify more than 95% of the ring boundaries in conifers, and might mark some false rings. In diffuse porous species, most of the ring boundaries are correctly identified, but a number of false rings could also be marked.

User intervention is required (through a graphical interface) to verify and adjust the ring identifications made by the program. This intervention is not time-consuming, and typically requires less than 30 seconds per sample. Samples can also be visually cross-correlated on screen to further assist in resolving ring identification conflicts. The identification and marking of "missing" rings (common in many slow-growing species) is also possible. The entire analysis is stored in a compact format (Internal Scan [ISC] format) that can be archived and, if necessary, reanalyzed at a later date without having to repeat the X-ray and scanning procedure.

DendroScan can produce reports of various ring parameters, including widths and densities for the entire ring, or the earlywood/latewood components. Such parameters are difficult and time-consuming to measure without digital analysis. Report formats include those required for the International Tree-Ring Data Bank Library (to facilitate development of standard chronologies) and an ASCII format, which is easily transported to spreadsheets for further analysis.

## 1.6 Time and Material Costs

Time costs depend on the nature and number of samples processed in a given batch. One 37.5-h work week is sufficient to prepare, X-ray, and analyze about 150 increment cores. This represents about 15 min per sample. (Note that not all of this time is operator time; a significant portion is consumed in waiting for extraction, pressing, etc., during which the operator can be doing other things.) A breakdown of the operator time per sample for each stage of the process is given in Appendix 1.

Operating costs are minimal. The X-ray film costs about $4.00 per 14- × 17-in. (35.6- × 43.2-cm) page, which can be used to X-ray several (up to 100) samples. The cost of photographic paper is about $0.40 per 8- × 10-in. (20.3- × 25.4-cm) sheet or about $4.00 per 100 samples. Including film, paper, and chemicals for processing, blade wear, shrink-wrap, and miscellaneous expenses, the total material cost is estimated at less than $0.50 per sample.

The system can be duplicated inexpensively. An X-ray machine similar to that used by the authors (the most expensive single component) can be purchased for less than $15,000; less-expensive machines are available. Suitable dark-room facilities can be assembled for less than $3,500 including tanks, safe-lights, and enlarger. A suitable table saw can be purchased for as little as $500; the blades are $200/pair and the jig about $120. Scanners (600 dpi or possibly 1200 dpi) can be purchased for less than $1,500. Extraction equipment is available for under $1,000. A suitable computer, if not already available, can be purchased for less than $2,000. The only custom-made parts of the system are the curved steel plates used in the sample press (about $30 each), the machined saw blade separator (custom-machined for about $100), the curved aluminium X-ray sample plates ($5.00 each if custom-made in lots of 50 or more), and the calibration wedge (which is included with this manual). Full capital cost estimates (replacement cost) are provided in Appendix 1. The total capital cost comes to under $27,000.

If only ring widths are required, the capital costs are reduced to the price of a computer, scanner, and sanding equipment (less than $4,000), and operating costs are somewhat lower.

# PART 2 SAMPLE PREPARATION

This section describes the apparatus and laboratory methods used to prepare samples for scanning and analysis. This includes collection, thin-slicing, and extraction, followed by X-raying and negative and image development. Each of these stages requires specialized techniques or equipment and some specialized facilities such as extraction tubes, fume hoods, and drying ovens.

## 2.1 Sample Collection

The sample collection procedures for X-ray analysis are more exacting than those for traditional visual analysis. Two types of samples are most commonly used: disks and increment cores. Disk sampling requires the destruction of a tree, and disks are expensive to acquire and transport. They are easy to prepare for X-ray, however, and allow the analyst to select blemish-free radii for analysis. Increment cores are easier to acquire, although they often miss the pith of the tree and require more-intensive preparation for X-ray.

Whether disks or cores are taken, some measurements of the tree should also be taken. Measuring the stem diameter and bark thickness at the sampled locations is particularly important. This ensures that final ring widths can be corrected to green dimensions during data analysis. The height of the tree should also be measured if incremental biomass and volume growth rates are to be later calculated from the rings.

### 2.1.a Increment Cores

The manual *Increment Core Sampling Techniques for High Quality Cores* (Jozsa 1988) provides many useful hints for collecting good samples; the most important is to keep the cutting edge of the increment borer sharp. A dull edge frequently results in twisting and breakage of cores. Also important is cutting the core at a 90° angle to the bole of the tree, ensuring that the growth rings pass through the core as a plane perpendicular to the longitudinal axis of the core. This is usually easy to accomplish, but can be complicated if cores are taken from the base of the tree, or if the stem is twisted and leaning. If time and resources allow, more than one core should be taken, preferably with cores 180° apart from each other. This maximizes the probability of striking the pith and allows an accurate estimate of the density of the tree (Gonzalez 1989). The cores should be stored in unwaxed paper or plastic straws with the ends stapled or crimped (not taped) so that moisture can escape. This allows the cores to dry in the straws and limits decomposition.

### 2.1.b. Disks (Cookies)

Disks, also referred to as cookies, should be about 5-cm (2-in.) thick. If the disks are thicker, they tend to develop multiple microscopic fractures as they dry, causing the final X-ray slices to fragment. If the disks are too thick or too thin, they become too cumbersome or too small for safely cutting off a subsample for X-ray using standard woodworking tools. The disk should be cut at 90° to the longitudinal axis of the bole at the area being sampled. Furthermore, both cuts should be parallel to ensure that the disk is not wedge-shaped, as this complicates further processing.

## 2.2 Thin-slicing the Samples for X-ray

Samples of wood for X-ray densitometry must be thin enough that all structures (i.e., ring boundaries) can be treated as straight vertical walls. If the sample is too thick, there could be vertical curvature or tipping in some of these structures. Vertical curvature will degrade the precision of both the width and the density measurements. Conversely, overly thin samples are too fragile to work with effectively. Most X-ray densitometry systems work with samples 1–2 mm thick (e.g., Fritts 1976; Jozsa and Myronuk 1986).

The authors developed a low-technology, low-cost method: an ordinary table saw with twin blades and a modified jig. Increment cores are too flimsy to thin slice immediately; consequently, they must first be mounted securely in a block of wood. Disks must also be prepared by reducing them to a manageable size for mounting into the jig.

### 2.2.a Mounting Increment Cores

Core samples are mounted tightly into slots in blocks of wood. The blocks are 4-cm wide × 2-cm thick × 20-cm long with two slots on each face (Fig. 2.2.1). The slots are 5.5-mm wide and 4.5-mm deep (to accommodate cores taken with a 4.5-mm or 5.5-mm barrel). The blocks can be easily made from nominal 1- × 2-in. (2.5- × 5.1-cm) fir boards, the slots cut with a dado or custom-made saw blade. Fir is preferred over spruce because it does not warp as readily and does not dull the blades as quickly.

The cores are mounted in the slots using a liberal amount of a protein-based, water-soluble glue and then pressed and dried in a 70°C drying oven. Often

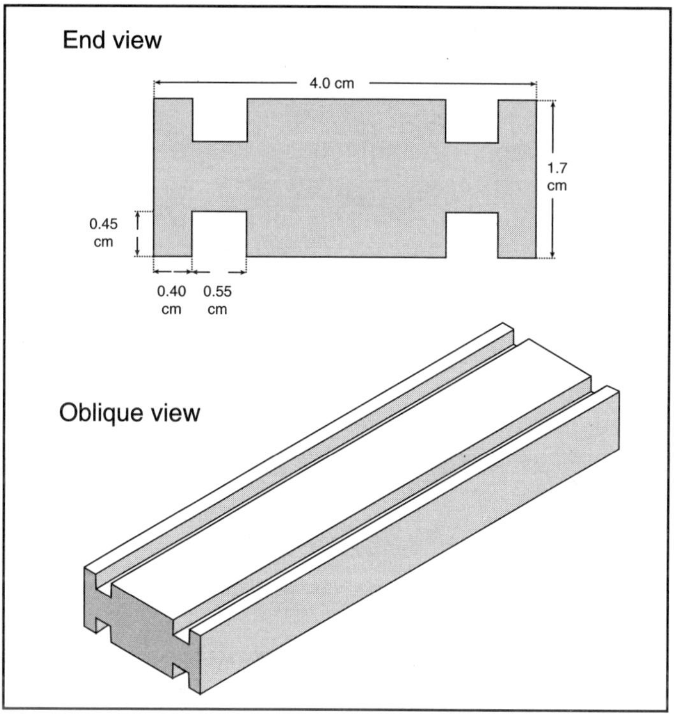


**Figure 2.2.1 Illustration of wood block for mounting increment cores.** These blocks can be constructed fairly cheaply from nominal 1 × 2 in. fir stock. The grooves can be dado-cut using an ordinary table saw.

the glue shrinks after the first application, leaving gaps that could result in breakage during the thin slicing. To remedy this, more glue can be applied and the samples air-dried.

The type of glue used for mounting increment cores has a significant effect upon the quality of the X-ray pictures. The glue needs to be strong enough to hold the core securely during thin slicing, but must not significantly penetrate into the wood or it will obscure the X-ray image. It was found that solvent-based glues invade the wood too readily; water-based carpenters' glues work better because they form a filter cake on the surface of the wood, restricting invasion of glue into the pores. Some white (aliphatic resin) carpenters' glues were found to be unsatisfactory as they appear to form a polymerized skin after curing and were resistant to later removal by water or solvents.

Protein-based glues such as Franklin Hide glue do not form this skin and appear to be completely removed by a water extraction.

The cores must be aligned with the grain in the appropriate direction. Often, due to discoloration or cutting burrs, it is difficult to determine the direction of the grain from the ring pattern alone. An examination of the surface of the core, though, will reveal two flat or shiny bands on opposite sides of the core. These bands occur where the coring drill blade split the fibers rather than cutting through them. The wood grain is always parallel to these bands, and they are the most reliable method of determining the direction of the grain in cores, especially in diffuse porous woods where details of the wood grain are often invisible. If the bands appear to spiral around the core, then the core is twisted and might not be useful for X-ray analysis. Figure 2.2.2 shows how cores should be aligned within the groove.

### 2.2.b Preparing Disks

Disks should be reduced (using a bandsaw) to wooden bars about 2-cm wide by 4-cm thick (0.75 × 1.5 in.), encompassing a diameter section and including the pith close to one edge.

### 2.2.c Using a Table Saw to Thin Slice

A regular 25-cm (10-in.) table saw run at 2500 rpm and two 25-cm (10-in.) carbide-tipped blades (Fig. 2.2.3) are used to slice the final X-ray sample. The blades should have at least 80 teeth to promote a smooth cut. Both cross-cut and triple-chip tooth patterns are acceptable. Triple-chip blades are better in that the tooth pattern stabilizes the blade and minimizes vibration. Blades are available in thick and thin bodies. Thick-bodied blades are preferable because they are more resistant to warping (a warp of even a fraction of a millimetre will cause many samples to shatter). The blades should be stabilized with two 12.5-cm (5-in.) diameter blade stiffeners and separated by a machined steel separator disk.

A jig mounted in the key slot of the table saw is used to hold the sample rigidly while cutting, further minimizing vibration and ensuring operator safety (Fig. 2.2.4). An inside cut is made about 6–10 mm into the sample blocks, leaving a subsample suspended between the two cuts from the double blade. This subsample is then removed using a razor knife and is ready for extraction.

In tests conducted by the authors with various types of wood, all but severely decomposed samples were found to be easily processed with minimal breakage. Air-dried samples were found to resist breakage better than fresh samples.

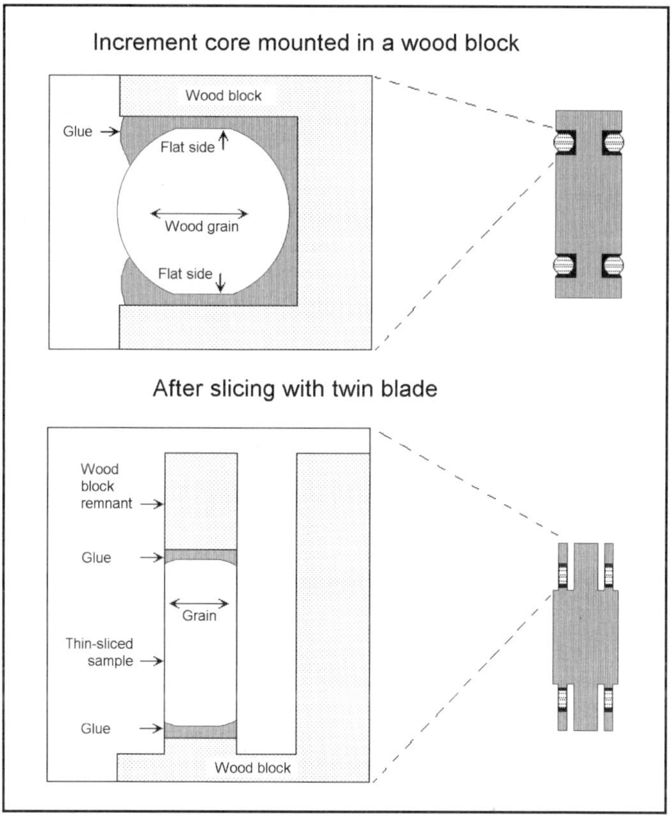


**Figure 2.2.2 Diagram of mounted increment core before and after slicing.** Note that the core is oriented using the flat sides/bands typical of a split grain surface. This ensures that thin-slicing occurs perpendicular to the wood grain, allowing for sharp detail in an X-ray image.

Preliminary tests showed that denser woods such as tamarack and black spruce could be sliced to as thin as 1 mm by using a thinner separator disc (thinner samples enhance the resolution, which is especially important in samples with rings thinner than 0.5 mm).

The following tips will ensure a repeatable, precise cut:

- Maintain sharp blades. Inspect their condition after every use and have them sharpened as necessary. Dull blades will result in ragged cuts and a high probability of shattering samples while slicing.

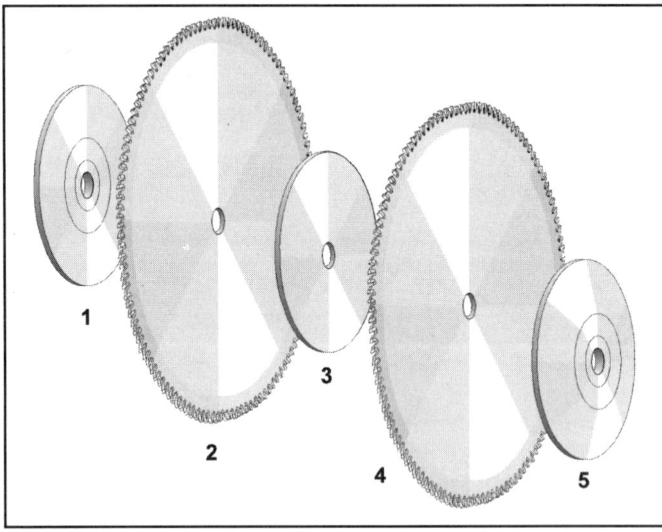


**Figure 2.2.3** **Table saw twin blade assembly.** The blade stiffeners (1 and 5) reduce vibration. The blades (2 and 4) have a minimum of 80 carbide teeth arranged in a triple-chip or cross-cut pattern, and are always mounted in the same order to maintain quality control. The separator disk (3) is precision-machined to a uniform thickness to minimize wobble.

- Keep the blades clean. The saw blades should be cleaned after every use with a 10% solution of potassium hydroxide (very corrosive). This solution will dissolve the resin that builds up on the blades during cutting. Make sure the blades are completely washed and dried after cleaning.

- Store the blades so that they do not contact one another or any other steel part, to avoid chipping of the carbide teeth.

- Always mount the blades in the same order on the saw. This is easily done if labels are engraved on the blades. This ensures that the cuts are repeatable.

- Make sure that the samples are clamped securely in the jig. If the samples move during the slicing operation, the quality of the slice will be severely degraded.

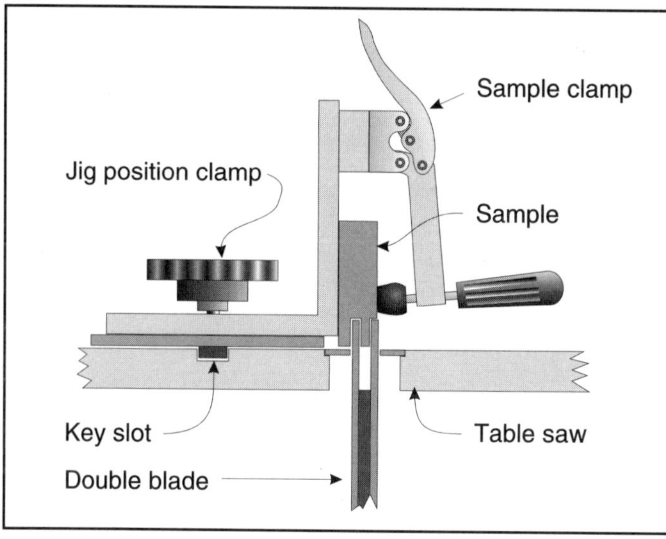


**Figure 2.2.4** **Table saw thin-slicing jig.** The actual design of the jig can vary depending on manufacturer. The key features are the attachment to the table saw using the key slot, and a secure sample clamp.

- Periodically check the sample thicknesses with a precision caliper to maintain quality control. This is essential after sharpening, as the sharpening process always affects the width of the saw kerf; and, consequently, the thickness of the samples.

During the cutting process, the original identification information is unavoidably removed from the samples and the subsamples need to be re-marked. It is recommended that the complete identification of the sample be written directly onto the subsamples immediately after cutting with a soft pencil, and renewed as necessary. Pen marks were found to be extracted too easily by organic solvents, whereas pencil marks both survived extraction and were not thick enough to obscure the X-ray image.

## 2.3 Extraction and Drying

Extraction of resins, salts, and glue present in the samples is recommended before the final samples are X-rayed (Jozsa and Myronuk 1986). The purpose of extraction is two-fold. Subfossil wood samples have often lost at least some of the extractives through natural processes, and are therefore poorly comparable to fresh wood unless extractives are completely removed from both. As well, extractives are not homogenously distributed within the wood; some (water-soluble resins) are concentrated in the sapwood and others (particularly oil soluble extractives) in the heartwood. While this distribution has allowed X-rays to be used for distinguishing the sapwood/heartwood boundary (e.g., Polge 1964), it interferes with measurement of wood density and obscures fine rings (Fig. 2.3.1). The water- and oil-soluble extractives should be removed for most studies except where tree health at the time of sampling is a factor for consideration.

Two extractions are recommended. A hot-water extraction should be conducted for 12–24 h to remove inorganic salts and water soluble resins. The organic solvent extraction (6–8 h) will remove aliphatic and aromatic resins with higher molecular weights. The solvent recommended is a mixture of 70% cyclohexane–30% ethanol (Josefina Gonzalez, April 21, 1991, FORINTEK Canada Corp., personal communication). Solvent extractions should be conducted under a fume hood and protective gear should be worn. Care should also be taken to allow no sources of ignition near the extraction apparatus, as the organic solvents are extremely flammable.

Fractured samples can become disarticulated and disordered during extraction. This can be minimized by first wrapping the samples into a roll of open-weave nylon cloth. This will allow many pieces to be extracted at once without risk of losing samples or sample fragments. Nylon is recommended because it is stable and has a low absorbency. Make sure that all sample fragments are appropriately identified with soft-pencil marks prior to extraction.

Water extraction can be conducted with fairly simple apparatus. One option is simply to boil the samples in a pot of water. Another is to use hot tap water coupled with a simplified extraction flask with siphon (Fig. 2.3.2). The latter method has the advantage of requiring little or no supervision over a long period of extraction.

Solvent extraction should be performed in a Soxhlet extraction apparatus (Fig. 2.3.2). A volume of solvent is placed in a boiling flask, which is heated. The vapor from this flask passes into a condenser and drips into a tube containing the samples. The samples soak in this liquid, which fills the tube and overflows through a siphon back into the boiling flask. This apparatus allows the samples to be extracted continuously with clean solvent, but

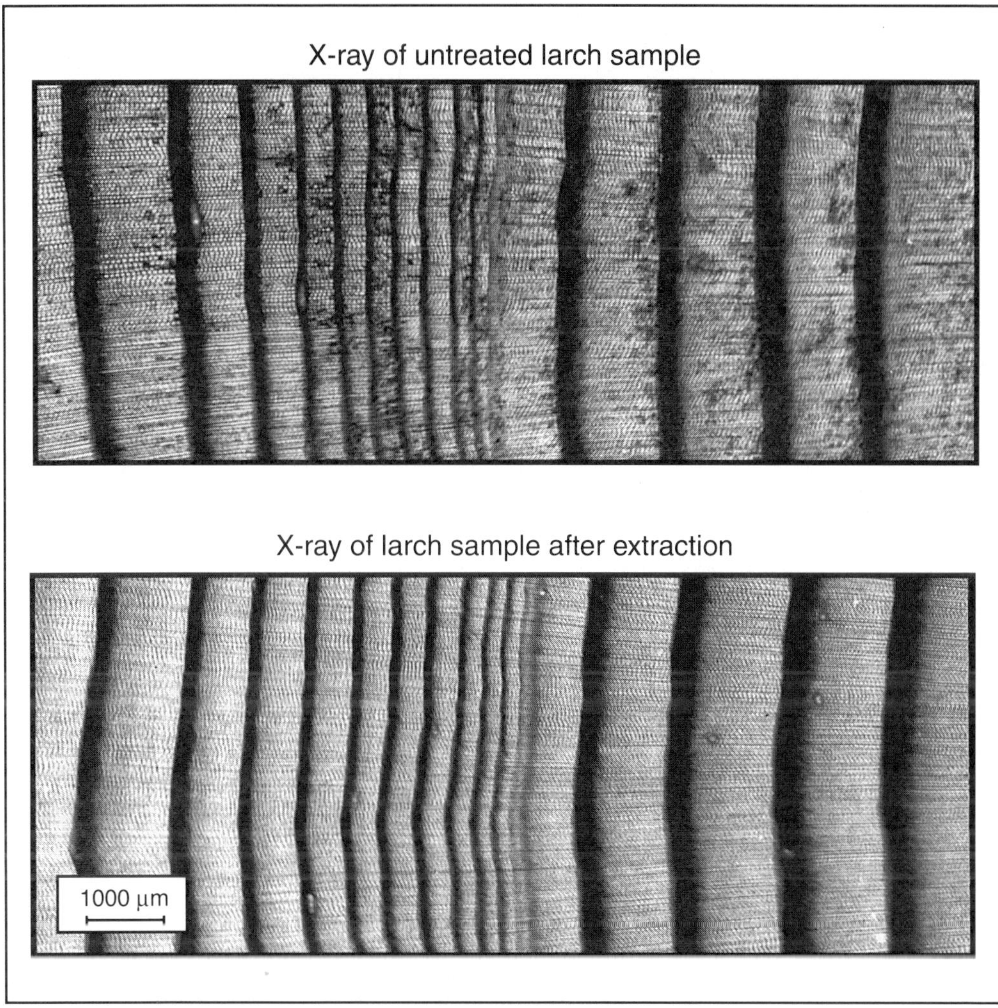


**Figure 2.3.1 Comparison of X-ray quality before and after extraction.** These samples are of eastern larch (*Larix laricina*) heartwood. Extraction was conducted with hot water for 12 h and 70:30 cyclohexane/ethanol for 8 h.

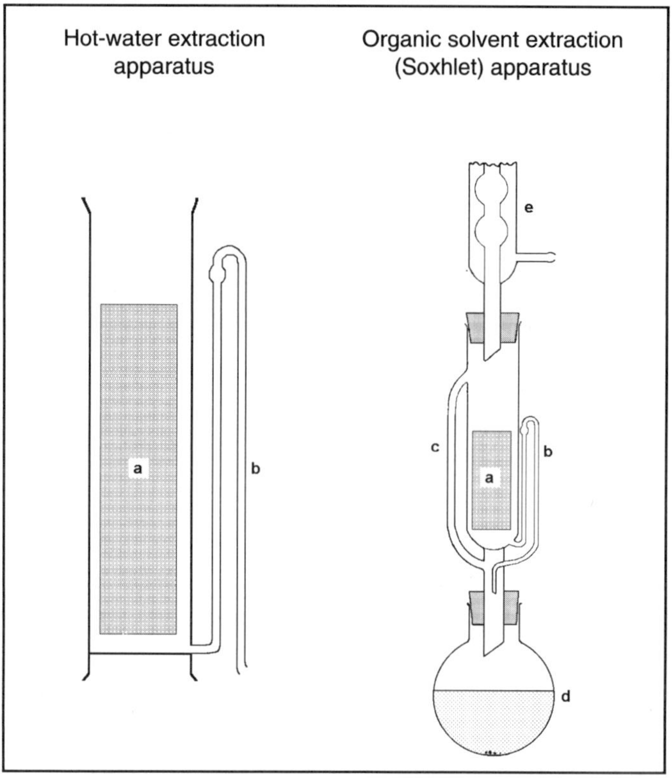


**Figure 2.3.2** **Extraction apparatuses.** The hot-water extraction flask is loaded with samples wrapped in nylon mesh (a). Hot tap water is run continuously into the top of the flask. The flask fills to the top of the siphon (b) and then empties. The solvent extractor is designed to minimize solvent loss. The solvent is boiled in a 1500-mL flask (d), evaporates through the evaporative arm in the Soxhlet flask (c), and condenses in a water-cooled condenser (e). Solvent thus drips onto the samples (a). When the flask fills to the height of the siphon (b), the solvent empties back into the boiling flask.

minimizes the quantity of solvent required (the solvent can also be distilled after extraction and largely recovered).

After extraction, the samples are placed between sheets of blotter paper, pressed between metal plates, and dried at 70°C. The drying process can take as long as 48 h. The press is made of slightly curved plates of cold-rolled steel about 6-mm thick, coated with a plastic paint to prevent rusting and sample contamination. The press at NoFC has a circular curvature equivalent to an arc with a radius of 625 mm to match the source-to-subject distance in the X-ray machine. This curvature is imposed lengthwise on the samples to eliminate parallax in the X-ray chamber (details are discussed in the next section).

## 2.4 X-raying and Film Development

The X-ray machine used at NoFC is an enclosed unit (TORREX 120, manufactured by Torr X-ray Corporation of Los Angeles) that was previously used for examining seeds. It is a table-top machine that runs on 120 volts alternating current (VAC), and has a variable transformer capable of producing 0–120 power kilo electron volts (PKeV); the X-ray tube is forced oil-cooled, and is rated at either 3 or 5 milliamps (mA) (push-button selected). The imaging chamber is fully shielded, and the door includes a safety interlock that prevents X-ray emission when the door is open. A constant voltage power supply is employed to maintain stable power to the X-ray machine. The distance between the X-ray source and the film plane is 625 mm. The X-ray machine and development system are all housed in a darkroom with X-ray safelights, to facilitate the handling of raw X-ray film.

AGFA D3-SC is an excellent X-ray film, with very fine grain and a single emulsion. It yields sharp images with high contrast and sufficient resolution that the cell walls can be resolved on the film. Any high resolution industrial X-ray film should be sufficient, but a single emulsion film is recommended to maintain a crisp resolution. Many X-ray films have an emulsion on both sides to reduce exposure time, but the two emulsions can result in parallax blurring of the image, both during irradiation and when used in an enlarger. The authors found it cost effective to purchase the film in 14 × 17 in. (35.6- × 43.2-cm) sheets, and then cut it to 4- × 5-in. (10.2- × 12.7-cm) sheets in the darkroom.

It is important that the darkroom be humidity-controlled (Schweingruber et al. 1978). Wood is very hygroscopic; oven-dried samples can gain 2–3%

moisture within 15 min of removal from a drying oven, even in an environment with relative humidity less than 20%. It is therefore impractical to X-ray the samples in an oven-dried condition. The best alternative is to allow the samples to equilibrate with a constant humidity atmosphere, and then apply a correction to the measured densities. The darkroom at NoFC is controlled with a humidifier at 30% relative humidity. At this rate, the moisture content of wood samples is 6%. It is recommended that the samples be allowed to equilibrate in the humidity-controlled darkroom for 4–8 h prior to X-ray. (Appendix 4 contains a graph showing the equilibrium moisture levels in wood for different relative humidities.)

The calibration wedge (Fig. 2.4.1) is included in the back cover of this manual. This wedge has been precalibrated, and is a fragile precision instrument. Avoid dropping the wedge on a hard surface or handling it in a rough manner. It is very brittle and the tip breaks easily. Store it in a place where it can be easily found under darkroom conditions. Appendix 4 describes how to re-calibrate a wedge if the tip has been slightly damaged.

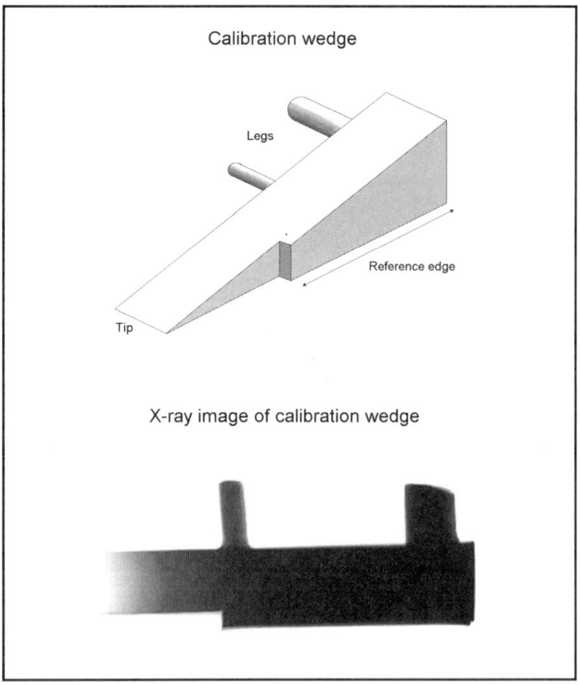


**Figure 2.4.1 X-ray calibration wedge.** The lower diagram illustrates how the wedge appears on an X-ray.

The samples and a calibration wedge are arranged on a thin acetate film (same dimensions as the X-ray film) on aluminium plates with a curvature matching the source-to-subject distance in the X-ray chamber. The calibration wedge is always placed as the topmost item in the image with the legs pointing toward the top of the image and the tip pointing to the left (Fig. 2.4.2). Samples are placed with the bark end to the left, one above the other such that the radii to be analyzed are horizontal on the plate. Samples that are too long to fit on the image are broken into shorter pieces and stacked as illustrated in Figure 2.4.2. The first piece should be placed in a normal position, but the other subsamples should be indented as indicated. This makes it easier to identify them as subsamples on the final image.

After the samples and wedge have been arranged on the acetate, the plate is taken into the darkroom and the raw film is slid underneath (emulsion side up). Tape or lead weights are used to secure and contour the film and acetate to the plate. A typical exposure for a 1.4-mm-thick sample is 3.00 min at 3 mA and 12 PKeV.

The acetate does slightly reduce the penetration of X-rays to the film; however, because the acetate is very thin, and both samples and calibration wedge are located above it, the effect will be cancelled when the image is analyzed using DendroScan. The use of the acetate allows the fragile samples to be arranged under full light and remain undisturbed when the film is added in the darkroom. The alternative is the difficult task of arranging the small samples directly on the X-ray film under darkroom lighting.

The film can be developed manually using a three-tank system in a 20°C constant-temperature bath. The films are arranged in groups of four in developing frames designed for handling 4 × 5-in. films/prints. The tanks are large enough to contain three frames, allowing a total of 12 films to be developed at once. Development times are as per manufacturer recommendations: 5 min development, 1-min stop bath, 5 min fixer, and a 20-min water wash. With suitable planning and preparation, up to 40 images (about 500 samples) can be X-rayed and developed in approximately 4 h. Tray development systems can also be used, although this will restrict the number of films that can be developed per batch. Tanks can be built from PVC or ABS plastic sheets for a reasonable price and are well worth the investment. Floating covers in the baths reduce evaporation and oxidation of the liquid chemicals between use, prolonging their effective lives.

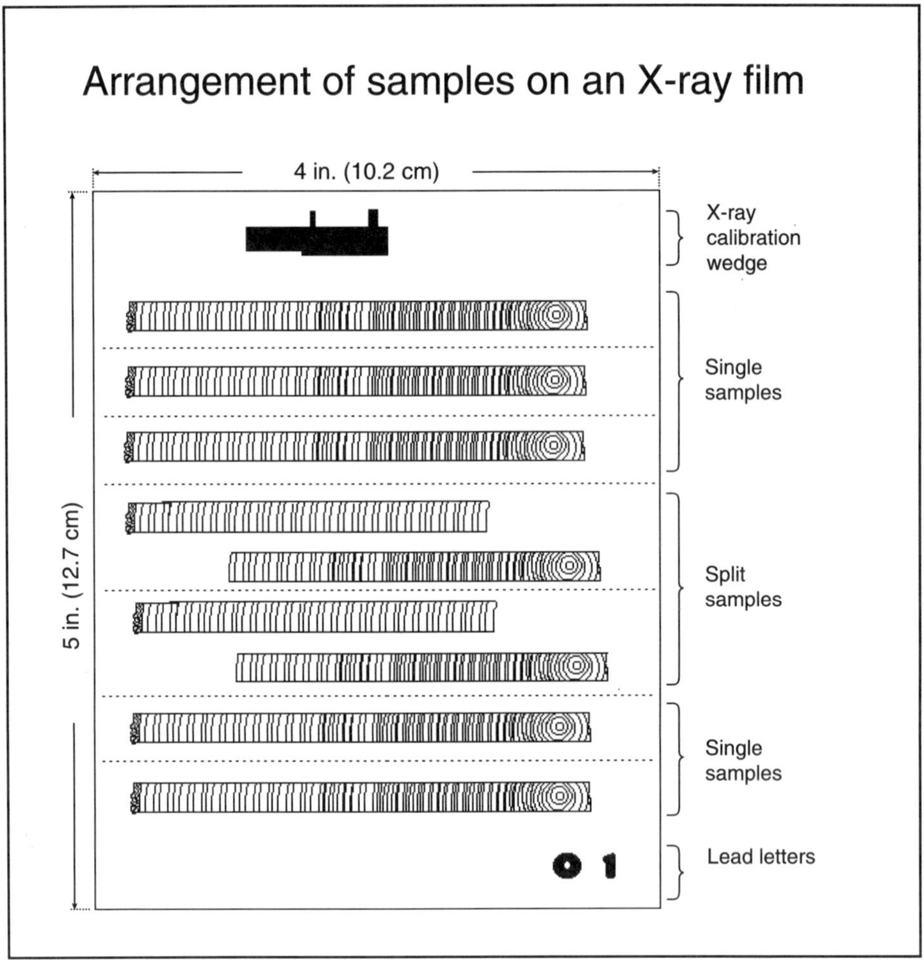


**Figure 2.4.2 Samples and wedge arranged on an X-ray film.** The wedge must be the first item at the top of the image. Samples less than 10 cm in length can be placed horizontally. Samples longer than 10 cm can be split, with the second piece (and third or fourth pieces) indented to facilitate their later identification on the X-ray image. Lead letters used to identify the image should be placed as far as possible from the samples or wedge to prevent interference. The entire arrangement is placed on a 4 × 5 in. (10.2 × 12.7 cm) piece of thin acetate.

## 2.5 Print Development

The negatives can be scanned directly if a transmission scanner is available, though such scanners with a reasonable resolution and image area tend to be quite expensive. A lower-cost alternative is to produce positive enlargements, and scan these enlargements. The authors have found that high quality black-and-white enlargements can be made with a standard enlarger (most of which can handle large format 4 × 5 in. films). Image quality can be variable depending on lighting conditions and the condition of the negative. High resolution and good contrast can be achieved with the following recommendations:

- On a regular basis, use a light meter to confirm that the brightness of the enlarger light is consistent throughout the frame when no negative is present in the light path. Most enlargers are designed to give equal brightness throughout, but this can be radically affected by poor bulb placement, dust on the lenses, etc. Clean and adjust the enlarger as necessary.

- Use a glass mount film holder that securely sandwiches the film between two plates of glass. If an edge-mount film holder is used, heat-related deformation and movement will cause blurring.

- Use the smallest aperture (largest f-stop number) available. This will result in longer exposure times but superior image detail. Exposures of 2–5 min with an aperture of f-45 are feasible. The smaller aperture increases the depth of field of the enlarger system, resulting in sharper focus throughout the image.

- Use glossy multi-contrast paper. The flat surface on glossy paper results in finer detail than is possible with mat papers. The high reflectiveness of the paper surface is not a problem when scanning, as the scanner detectors are placed at an angle that avoids surface reflections.

- Use a low-contrast filter (00 yields the best results). This reduces the amount of bleed between lighter and darker areas of the paper.

- The final image should have sufficient detail that larger wood features such as resin ducts are clearly visible. The lightest portions of the earlywood should be dark enough that they will not be confused with the background. The darkest portions of the latewood should be dark, but not necessarily jet-black. The objective is to ensure that the gray values in the samples fall in the workable range for the scanner.

If the above points are followed, the resulting image will have good contrast and contain detail with a resolution of about 0.01 mm. This will allow the image to be scanned at up to 1200 dpi without loss of resolution due to poor image quality.

Note: No matter how carefully the enlargements are made, the photograph will always have poorer resolution than the X-ray film. Cellular detail is usually not visible in prints (even if they are clearly visible in the negatives), although larger details such as resin ducts are usually readily apparent.

## 2.6 Preparation of Wood Samples for Direct Scanning (Visual Analysis)

DendroScan was designed to analyze X-ray samples, but it also has the capability of analyzing images of directly scanned wood surfaces as well. Widths of annual rings, as well as widths of earlywood and latewood components, can be measured and reported. It is not possible, however, to analyze wood density in this manner.

Samples (cookie or core) must be surface-polished by sanding to at least the level of 400-grit paper. This is a time-consuming process, as it involves starting with a coarse grit of sandpaper (50 or 100, depending on the condition of the original surface) and proceeding to successively finer grits. The objective is to achieve a final surface with no scratches, where the wood element cell walls are clearly visible. Disk and belt sanders can be used for the coarser sanding, but the final polish is best-achieved with a hand-held orbital sander.

Sanding of cores can be made easier if several are glued side-by-side into shallow grooves in a piece of plywood. Between 20 and 50 samples can thus be sanded together, significantly reducing the operator time required for a good finish. The sanded surface must be perpendicular to the longitudinal axis of the tree (perpendicular to the wood fibers) to achieve best results. It is helpful to use the two flat/shiny bands on either side of the core to align the samples.

Disks (cookies) can require considerable work if the entire surface is sanded. This workload can be reduced if only one or two radii need to be analyzed.

The disk can be reduced to a strip running across the face and including the pith using a band saw. Several strips can be mounted side by side and sanded together as if they were cores. This not only reduces the sanding time, but also reduces the scanning and analysis time because several samples can fit on one scanned image.

For visual analysis, do not write the sample number directly on the surface of the wood, as it will interfere with the image analysis. The only limitation with visual analysis will occur if the tree rings are too narrow to be resolved by the scanner. This can be overcome by taking a close-up photograph of the samples (including a scale bar) with a macro-lens and producing black-and-white enlargements. The photographs can later be scanned to yield more precise measurements of tree rings. Enlargements can also be made on a photocopy machine, although the results will depend largely on the initial clarity of the rings and the quality and resolution of the photocopy machine.

# PART 3 SCANNING IMAGES AND MARKING SAMPLES

Part 3 describes the rules for using scanning software to draw digital black lines on scanned images. DendroScan reads information directly from an image file. This allows it to be functional with a wide range of scanners and imaging hardware. Various scanners can therefore be used, but they must be capable of producing 256 gray scale images.

## 3.1 Scanners, Software, and Image Resolution

Most scanners are bundled with scanning software capable of producing image files in a 256 gray scale Tagged Image File (TIF) format. Usually, the scanning software is also capable of some image editing and can be used to draw a series of black lines on the digital image. Scanned images do not usually contain absolute black unless the contrast has been enhanced; thus, lines drawn in black are distinguishable from the image itself, and can be used to define paths that DendroScan can follow to survey the image.

Note: TIF image formats are very versatile, and they occur in thousands of variations. DendroScan was designed to handle several variations of the 256 gray scale TIF format to be compatible with most scanning software. The authors have tested compatibility with several packages, including Microtek Imagestar, Adobe Photoshop, Corel PhotoPaint, and LView Pro. Not all scanning software, however, will necessarily produce files that DendroScan can read. The user should test compatibility with any proposed scanner software. If problems with compatibility arise, check the structure of the TIF file using TIFCHECK (*see* Appendix 5).[1]

When shopping for a scanner, it is important to note that the advertised scanning resolution is sometimes overrated. The optical resolution is the physical sampling rate conducted by the scanner (which might be different in the horizontal and vertical directions). Scanners often try to increase the apparent scanning resolution by mathematically interpolating between pixels. The advertised scanning resolution is often the maximum interpolated

1 Users experiencing problems might wish to contact the authors.

resolution. Scanning for digital analysis should not exceed the optical resolution because interpolation doesn't actually yield new data. It is therefore important to examine the technical specifications first; if the optical resolution is not specified, the scanner should not be used for digital analysis. The authors use a Microtek 600G scanner with an optical resolution of 600 × 600 dpi (0.04-mm sampling resolution). With a 2× photographic enlargement, this yields an actual sampling resolution of 0.02 mm. Higher optical-resolution scanners are also available.

## 3.2 How DendroScan Analyzes a TIF File

This subsection describes the important rules for drawing digital black lines on an image to describe an analytical path. These lines are actually a graphical set of instructions that can tell DendroScan the location of samples, into how many pieces the sample is broken, and the angle of the rings.

### 3.2.a Locating Digital Paths

TIF files can be quite large (for example, a 600-dpi image 20 × 25 cm will occupy almost 30 megabytes (MB) of disk space. It is therefore not possible for a computer program to load the entire file into memory at once. DendroScan overcomes this limitation by reading only small portions of the file at a time. Figure 3.2.1 illustrates how, by reading one scanned row at a time, DendroScan can assemble all the pixels in a digitally drawn path. On the first pass, it reads one scanning row at a time, looking for black pixels, and recording Y coordinates of these pixels on a linear array. After encountering a line containing no black pixels, the program knows it has found a sample.

Samples can contain several individual segments, usually running slightly diagonally to remain perpendicular to the tree rings. DendroScan will recognize different line segments as belonging to one sample as long as it doesn't encounter an empty row between them (Fig. 3.2.2). There must, however, be at least one blank column separating the line segments. This blank column functionally becomes a gap in the linear array that the program uses to locate pixels. The program uses this gap to determine that two segments are independent. If the gap is absent, it will seriously affect how the paths are translated (Section 3.2.b).

The program might not retain all pixels, because of the method used to gather pixel locations. If two black pixels are located one directly above another, the lower pixel location overwrites the first. Thus, a vertical line drawn directly

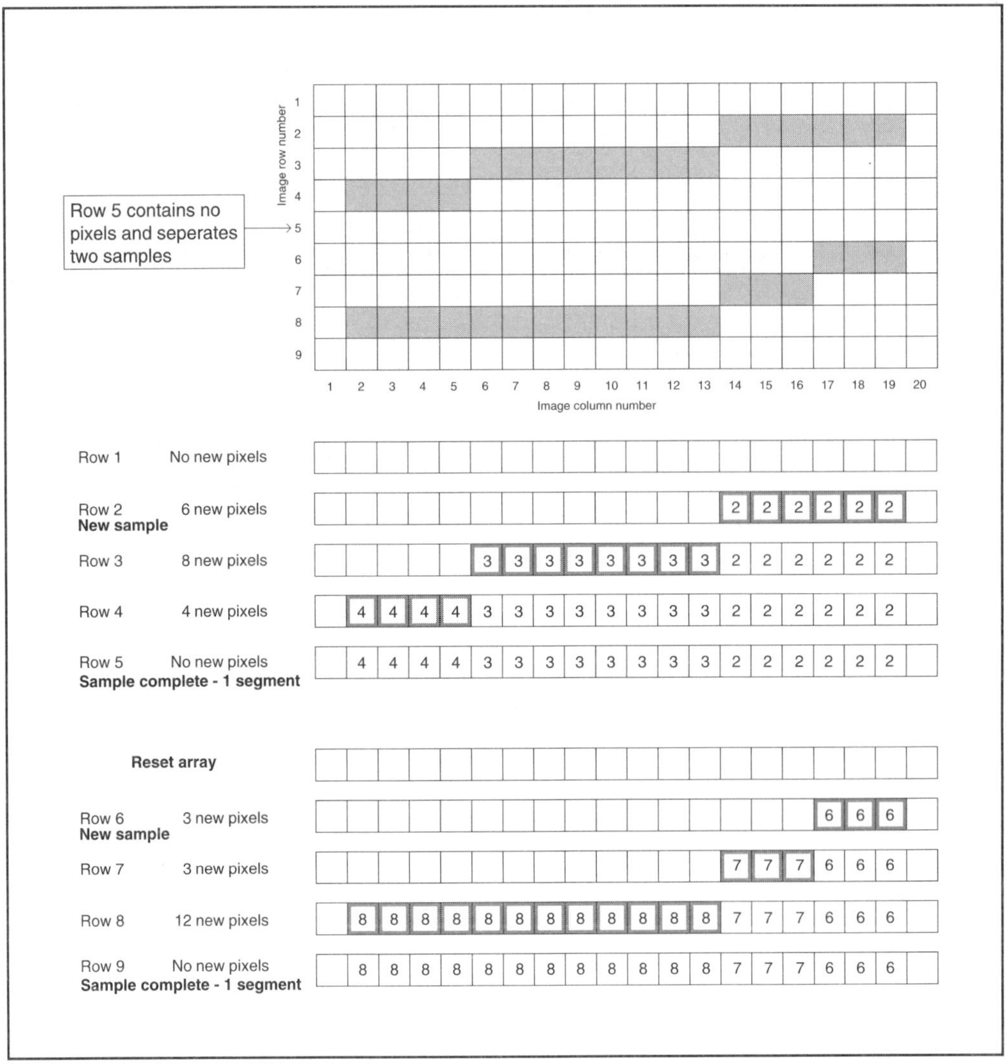


**Figure 3.2.1** **How DendroScan locates digitally drawn paths.** DendroScan uses a linear array (the same width as the image) to gather pixel locations. It scans the image one row at a time, and whenever it encounters a black pixel (shaded boxes in the image at the top) it records the row number in the corresponding array column position. When it encounters a row containing no black pixels, it processes the sample path and then resets the linear array. The routine repeats itself for the next sample(s) until it locates the end of the image.

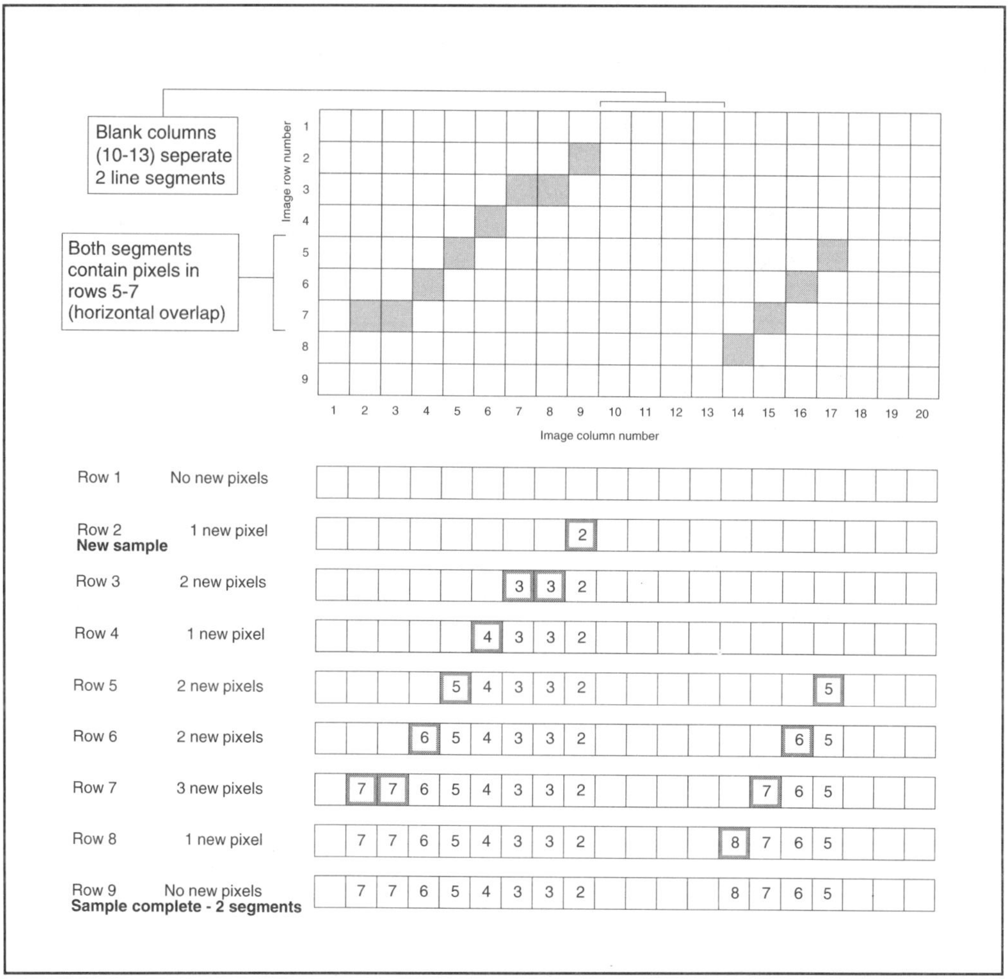


**Figure 3.2.2 How DendroScan locates a sample with multiple segments.** By scanning with a linear array, DendroScan will automatically identify multiple segments within a sample if they overlap horizontally (contain pixels in the same row) and are separated by a blank column. When a blank row is finally located (denoting the end of the sample), DendroScan can determine the number of sample segments using the gaps in the linear array.

above a sample path would be functionally invisible (this is called an "elevator line"). Such lines cannot therefore be used to define a path. They can be used in other ways. Figure 3.2.3 shows how elevator lines can instruct DendroScan to include two line segments as part of one sample, even though they may have a blank row between them. When the linear array is complete (when a blank row is encountered), all traces of the elevator line have disappeared to be replaced with the second line depicted. The program recognizes that they both belong to one sample, however, because the program does not encounter a blank row between the two segments.

### 3.2.b Translating the Digital Path

Finding the paths is only the first step conducted by DendroScan. It then needs to translate these jagged groups of pixels into smooth curves defining a logical analytical path that can be sampled at regular intervals. Figure 3.2.4 illustrates the basics of this method. First, a vector equation of the line is determined at each point using a 20-pixel average slope (in the illustration, the digital path is composed of 19 pixels, so by definition, the slope at each pixel is the same). A curve is then constructed by lining up the vectors end to end. New real-number coordinates are then calculated by subdividing the curve into equal widths based on the scanning resolution. These real-number coordinates become the functional sample points, as they are evenly spaced across the sample regardless of the angle of the digital path.

At each coordinate, a perpendicular transect is derived and a number of evenly spaced subsample points are laid out (the maximum number of sample points is user-selectable in the DendroScan initialization file). Given that the sample points are represented by real numbers, the density for each point is determined from the pixel whose center is closest to that point. Any black pixels are ignored because they do not represent image data. The remaining values are averaged to calculate the density value for the specific coordinates on the main vector. As the user is expected to draw the digital paths at right angles to ring boundaries, this averaging is parallel to the ring boundaries and does not significantly degrade the intra-annual density signal. Rather, it produces the effect of measuring several parallel paths, which increases the accuracy and repeatability of the densities measured.

This translation is conducted independently for each line segment in a given sample. The individual density signals are then joined end-to-end to create one continuous signal.

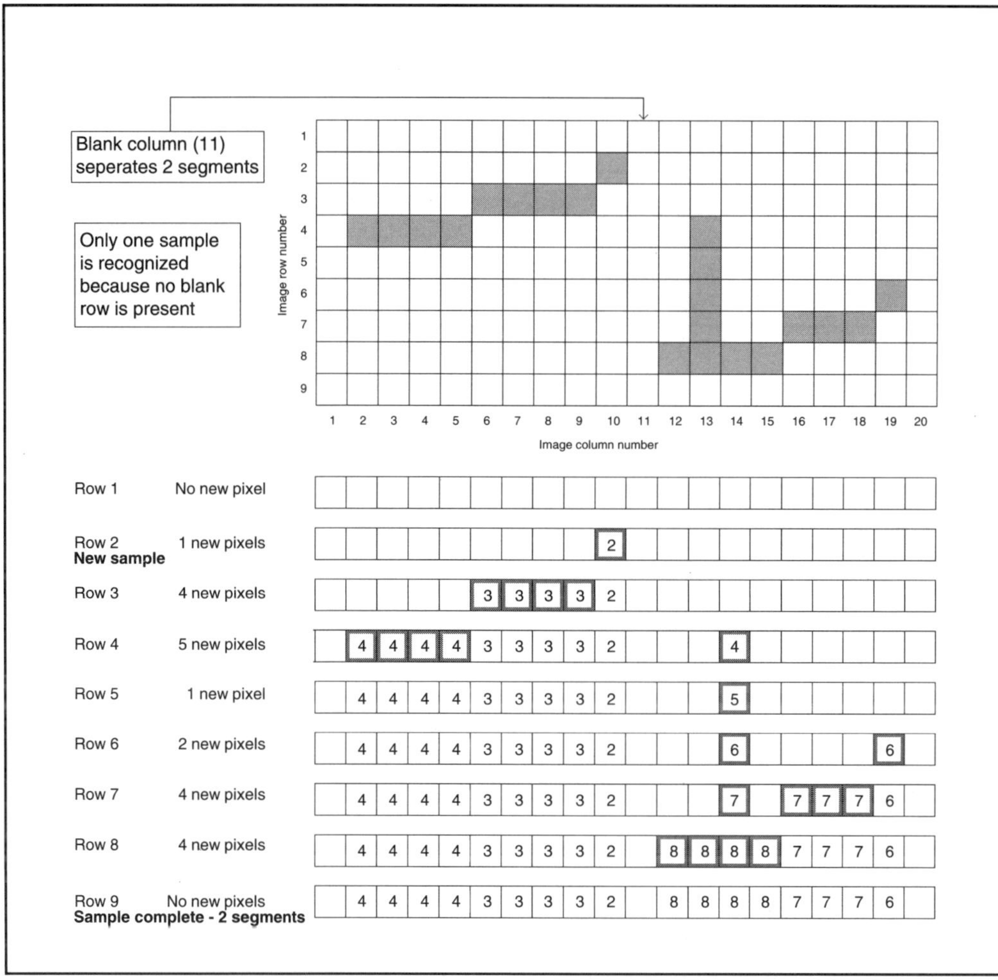


**Figure 3.2.3 How DendroScan locates sample segments linked with an elevator.** In this example, neither segment contains pixels on row 5; however, a line drawn vertically above the second sample (elevator line—see column 13) can prevent DendroScan from encountering a blank row until after the second segment. Although the elevator changes the definitions of the line segments, it does not itself define a path. As DendroScan scans downward, each pixel in the elevator line is replaced by the succeeding pixels found in the same column, rendering the elevator line funtionally invisible.

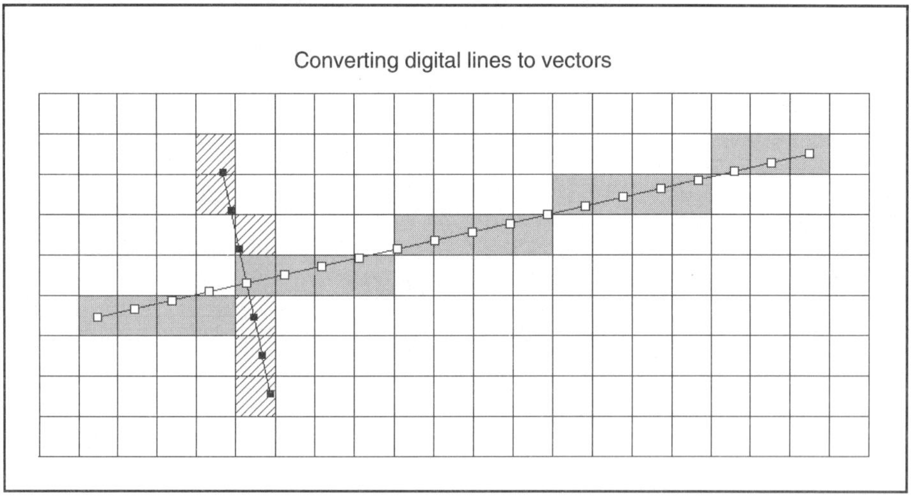


**Figure 3.2.4 How DendroScan reads densities along a digital path.** The gray squares depict the pixels in the digital path drawn by the user. DendroScan first calculates a vector (line superimposed on the digital path) using a 20-point average slope and then subdivides the vector into real-number coordinates (open boxes). The angular distance between the coordinates along the vector is equal to the image pixel size. This ensures that the image is sampled at regular intervals, regardless of the angle of the digital path, and that the width of these intervals does not exceed the resolution of the image. The image directly under the sample vector is usually occluded by the black pixels defining the digital path. As a result, the density value at any given coordinate is determined by averaging the densities of several pixels on either side of the main vector. At each coordinate, subsample points are evenly spaced along a transect perpendicular to the main vector (example shown passing through the fifth coordinate). The number of pixels used to calculate this average is user selectable in the DendroScan initialization file, and the distance between them is also one pixel width. The specific pixels used for the average (patterned squares) are those whose centers fall closest to the locations of these subsample points (black pixels are ignored). The actual algorithm includes additional measures to ensure that the sample vector doesn't drift from the digital path, but this simplified example summarizes the basic method.

Note: In Section 3.2.a, it was noted that independent segments within a sample should be separated by a blank vertical column. If there is no such gap between line segments, DendroScan will not recognize that they are independent. It will then derive a vector across the space between the two segments, and read densities across this space. This will functionally create a section of "garbage" data within the density signal, which will seriously impair later analysis.

## 3.3 Defining Analytical Paths by Drawing Lines

It is difficult to give detailed instructions on the tools to use because different scanners use different control programs. It is assumed that the user will first be familiar with the scanner software used before attempting to prepare files for DendroScan. The following guidelines should be followed:

- If scanning a positive image or photograph, set the scanner brightness control to slightly positive (+1% to +5%) to be absolutely certain that no true black pixels are measured in the sample image. Some scanners do not need this adjustment. The objective is to make sure that the lines later drawn by the user are the only true black pixels on the image.

- If scanning a negative, either set the scanner settings to produce reverse imaging, or simply scan the negative as-is and then reverse the entire image. Do not attempt to draw path lines until the image has been converted to a positive (dark means more dense, light means less dense). Users might need to experiment with the scanner brightness settings so that no black pixels remain on the image after it has been translated to a positive image.

- Make sure that the scanning resolution does not exceed the optical resolution of the scanner.

- Use a screen zoom ratio of 1:1 (also called 100%) when drawing lines. It is important that the user can see each individual pixel in the lines drawn.

- Use a line-drawing tool that produces one-pixel-wide black continuous lines. Be certain that the color of the line is absolute black, the darkest color available on the palette.

- Note that some programs such as Adobe Photoshop also have a "soft line" tool that produces a mix of grays to simulate a photographic line. Do not use this type of line tool. Not only does it produce a discontinuous line, but it adds gray pixels that DendroScan will read as density data.

- If the contrast or brightness is adjusted while lines are drawn (this is sometimes necessary when the user needs to see more detail) the user must make sure that these values are reset to zero before saving the file. Some programs will adjust the pixel values during a save if these values have been adjusted. The result could be that the black lines are no longer black and will therefore be invisible to DendroScan.

When an X-ray is being scanned, the first line drawn should be along the reference edge of the calibration wedge (Fig. 3.3.1). Ensure that this line is drawn consistently from image to image. DendroScan uses this line to locate the wedge and to determine the resolution of the image. Note that the tip of the wedge will often white out (disappear). This is not a problem, and will not impair the calculation of density values. The wedge is designed to handle a wide range of image conditions and is expected to reach gray-value saturation at either end. If no wedge is included (visual analysis), the program will assume that the image is not an X-ray. It will not attempt to calculate density but will analyze the ring boundaries using the image gray scale.

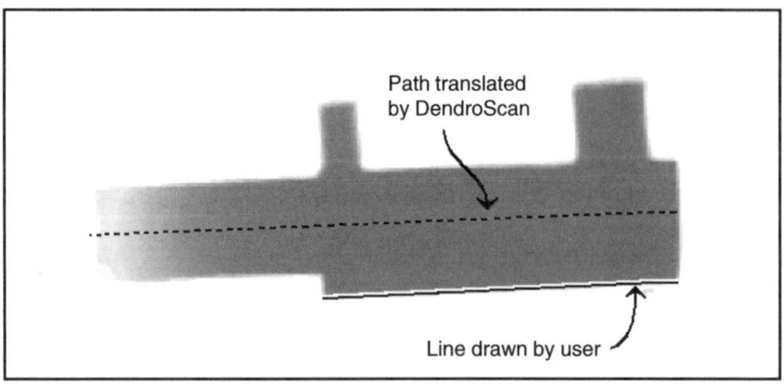


**Figure 3.3.1 Illustration of line-drawing on a calibration wedge.** The line should be drawn along the reference edge from the notch to the back, as shown by the solid line. DendroScan will use this line to determine the image resolution, and then will read the wedge along the path denoted by the dotted line, and calibrate the gray values in the image to density equivalents.

Black lines should be drawn on the samples, starting with the inside edge of the bark and perpendicular to the rings. In order to remain perpendicular to the rings and avoid irregularities such as cracks, resin ducts, and branch intrusions, the lines might have to curve slightly or be broken into different segments (*see* Fig. 3.3.2). This is feasible so long as the following rules are followed:

- Path lines should be drawn perpendicular to the rings. The closer to perpendicular, the more precise the density measurements become, particularly at ring boundaries where the density can change rapidly over a short space.

- Path lines can be drawn at any angle except true vertical. If a line is drawn vertically, DendroScan will only see the lowest point on that line and will not know how to deal with the line. (The program could crash as it attempts to determine the slope of a single point.)

- Path lines within a sample or subsample must horizontally overlap with other lines in the same sample/subsample.

- If two path lines do not naturally overlap horizontally, an elevator line can be used to tell DendroScan that the two lines belong to the same sample/subsample. An elevator line is drawn upward from the line lowest in the picture to a height that ensures horizontal overlap. The elevator line does not describe an analytical path because every point in the line lies vertically above an existing point in the line beneath it.

- A clear vertical gap must exist between line segments within one sample/subsample. If this vertical gap does not exist, DendroScan will translate an imaginary near-vertical path between the line segments where they start to overlap, inserting garbage data in the density signal.

- A clear horizontal gap extending the full width of the picture must exist between samples/subsamples. Without this gap, DendroScan will not differentiate among different samples.

- Lines should avoid blemishes and inclusions in the wood.

- Lines should not stray to the edge of the sample image. Many samples are slightly thinner at the edge, resulting in less-reliable density measurements.

- Wherever possible, start or end line segments at ring boundaries. This will yield more-precise measurements of ring widths. If the pith is missing, make sure the last line segment ends at a ring boundary.

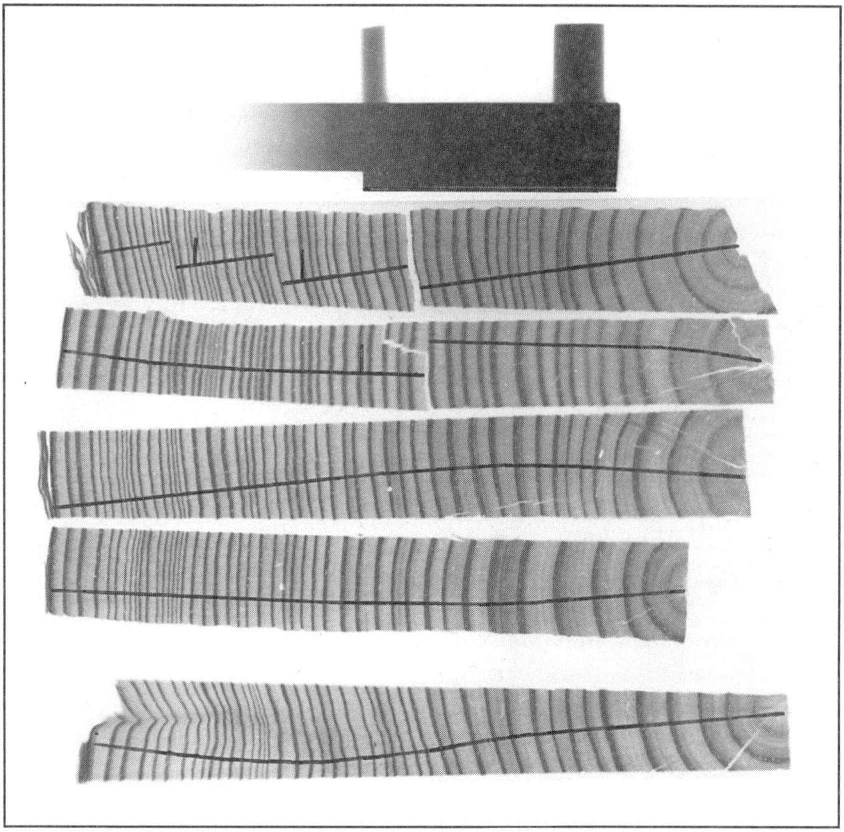

**Figure 3.3.2** **Example of path lines on an image.** In the first two samples, several line segments are used to accommodate the ring angle, and deal with cracks. Note that there is a slight vertical gap between every line segment, and that elevator lines are also used. The next two samples are less complex and single path lines were used. Note that they are slightly curved to remain perpendicular to the ring boundaries along the entire length. In the last sample, a stronger curvature was used so that the branch intrusion (deformed rings) could be avoided.

# PART 4 DendroScan USERS' MANUAL

DendroScan was developed over a period of 3 years, with the X-ray system and the program developed simultaneously. The program was initially designed to run on an IBM PC-AT or compatible (80286); however, memory requirements demanded that it be modified to operate on a faster machine. It will, nonetheless, run in a basic DOS environment within a memory space of as little as 450 kilobytes (kB).

With DendroScan, the initial task of scanning the samples and creating a TIF file must be done using the scanning software usually packaged with a digital scanner. DendroScan thus takes a specially prepared TIF image file of tree rings and extracts ring-width and density information. If a calibration wedge is included in the image, the conversion of gray-scale to density is performed automatically. If no calibration wedge is included, image gray values are recorded directly.

DendroScan translates the paths (black lines) predrawn by the user into image vectors. It then travels through the image along these vectors, analyzing the gray values or density values at intervals equivalent to the width of one image pixel. To reduce the variability in the values, it averages several pixels along a line perpendicular to the vectors. The number of pixels included in this average is defined by the user in the initialization file. The result of this analysis is an array of evenly spaced density or gray values. The density arrays are stored in a DendroScan binary storage format (ISC) after the samples have been located and preliminary analysis is complete. These files are typically 200 times smaller than the original TIF file from which they were derived. This allows large volumes of untreated data to be stored and transported on floppy disks; typically, about 100 samples will fit on one 1.44-MB floppy disk. The program can handle individual samples with up to 30 000 density values, corresponding to 127 cm (50 in.) if scanned at 600 dpi. Higher scanning resolutions will reduce the maximum sample size proportionately.

After the entire sample has been recorded, the program counts the rings using a multipass algorithm that recognizes both size and shape of ring signals. An image is produced on the screen, including a gray-scale image of the rings, a graph of the density along the radius, the estimated age of the sample in years, the direction of the bark, and those density peaks that were identified as rings by DendroScan. The operator might examine the marked ring boundaries at this point and edit the identifications made by the program.

Once the operator is satisfied that the rings are correctly identified, reports and graphs can be produced to display values such as ring width, summer maximum density, spring minimum density, average ring density, widths of earlywood and latewood.

The program has been used with both contact prints and enlargements from X-ray negatives, as well as with directly scanned samples of wood, microphotographs of thin sections of polar bear teeth, and both photographs and X-rays of laminated lake sediments. In all these cases, the program was able to detect and measure growth rings or laminations.

This manual assumes the reader is familiar with the fundamentals of DOS, including paths and common DOS commands. It also assumes that the user has access to a scanner capable of 256 gray shades at a minimum resolution of 300 dpi (also called ppi [pixels per inch]), and software that controls the scanner and creates/edits 8-bit gray scale files in the TIF format.

## 4.1 DendroScan Basics

### 4.1.a System Requirements

The following is required as a minimum for a practical system to run DendroScan:

| | |
|---|---|
| System | IBM or compatible PC with an 80286 processor or higher (80486DX or higher is strongly recommended). |
| Memory | 8 MB are recommended. This is not a limitation imposed by DendroScan, but rather the scanner, which usually requires Windows 3.x to operate the scanning software. If Windows 95 is being used, more memory might be required. (DendroScan only requires 450 kB of DOS memory to operate.) |
| Hard disk | A large hard disk is recommended because scanned images are large files (a single 8 × 10-in. image scanned at 600 dpi can take up 30 MB of disk space). A minimum of 200 MB of free space should be available for efficient operation of the scanner and DendroScan. Fortunately, the scanned images usually do not need to be kept after |

the initial analysis, and ISC files are much more compact than the images from which they were derived.

| | |
|---|---|
| Operating system | DOS 3.3 or higher is required for running DendroScan. Windows 3.1 is usually required for running a scanner. DendroScan will also run efficiently within a DOS application window. |
| Monitor | A VGA screen capable of displaying 16 colors is required for DendroScan. For running the scanner, 256 colors are recommended. |

### 4.1.b Installing DendroScan

Installation of DendroScan is simple and straightforward and is accomplished with the following steps:

- Insert the floppy disk in Drive A
- Switch to drive A **A:**
- Start the install program **INSTALL**
- The program will ask several questions. In each case, striking the <ENTER> key will cause the default value to be used. If the other values are unknown, the defaults will be inserted into the DENDSCAN.INI file, and can be later changed with a text editor as described in Appendix 3. The serial number of the calibration wedge should be input correctly, as indicated on the package containing the wedge.

**INSTALLATION DIRECTORY? (Default is C:\DS45)**

Installing to the default directory is recommended. Otherwise, type in a complete directory path name including the drive designation.

**SERIAL NUMBER OF THE CALIBRATION WEDGE? (No default)**

The entry expected is the five-digit serial number written on the calibration wedge included at the back of this manual. This will create a reference entry in the DENDSCAN.INI file under wedge number 1. If you cannot supply the correct wedge number, a roughly calibrated wedge will be generated by the install program.

**USUAL SAMPLE THICKNESS (CENTIMETRES)? (Default is 0.15)**

If you are uncertain of the thickness of samples that will be used later, this value can be changed later (*see* Appendix 3). The value expected is a real number from 0.01 to 1.0.

**WOOD MOISTURE CONTENT (PERCENT)? (Default is 10.0)**

The expected response is an integer value. The Install program inserts this value in the DENDSCAN.INI file. The moisture content should be the equilibrium moisture content at the expected relative humidity of the dark room. If the relative humidity of the X-ray darkroom is not known, then the default can be assumed temporarily. In this case, the calibration constant will have to be later changed when the relative humidity is known (*see* Appendix 4).

After the questions have all been answered, the program will uncompress and copy the program files to the specified directory.

After installation is complete, the following files will exist in the DendroScan directory:

| | |
|---|---|
| TIFCHECK.EXE | A utility program that checks the structure of TIF image files. |
| DENDSCAN.INI | The file containing basic information including calibration wedge data, and several user-selectable parameters. The wedge data in this file are customized for the specific wedge shipped with this copy of the manual. This file can be edited using an ordinary text editor. (*See* Appendix 3.) |
| DS45.EXE | The executable code for DendroScan version 4.5. |
| DS45.ICO | The Windows "File Properties" menu can be used to install the icon. |
| DS45HLP.HLP | Help file for DendroScan version 4.5. This file can be edited (to add personal observations or translate the content into another language, for example). (*See* Appendix 5 for details.) |

| | |
|---|---|
| DS45RES.RES | Resource file used by DendroScan. Do not attempt to edit this file. |
| EGAVGA.BGI | VGA screen driver required by DendroScan. |
| FILESEL.INI | The initialization file for DendroScan's file search utility. It keeps track of the last drive and directory that was accessed from DendroScan. |
| FIXLST.EXE | A utility program used after either the species list or help files have been edited. |
| SPECIES.LST | List of standard International Tree-Ring Data Bank (ITRDB) species abbreviations, for reference only. |
| SPP.LST | Reduced list of standard ITRDB species abbreviations. This list contains primarily North American and European species. This file is used by DendroScan, whereas the other file, SPECIES.LST, is not. A new SPP.LST can be modified to include any subset of SPECIES.LST. (Details are included in Appendix 5.) |
| TESTDS.TIF | A low-resolution sample image file containing samples with path lines already drawn. |
| INSTALL.EXE | The installation program. |
| INSTALL.DB | Wedge calibration data base. |

### 4.1.c Terminology used in DendroScan

Oven-dry density — Oven-dry density is the density of wood (g/cc) after the sample has been dried in an oven. The water content is assumed to be zero but in fact can be as high as 0.1% to 0.2%, depending on the temperature of the oven and the ambient relative humidity. This is the standard value that DendroScan attempts to measure. This value is difficult to measure gravimetrically in small samples because such samples imbibe water from the air at a rapid rate immediately after removal from the oven. The value can be calculated, however, if the samples are allowed to equilibrate in a constant relative-humidity environment and then weighed. Equilibrium water content at various relative humidities are shown in Appendix 4.

**Basic density**

Basic density is a theoretical value defined as the oven-dry weight divided by the fresh volume. It is theoretical because the actual fresh volume can vary depending on the ambient environmental conditions when the sample was taken. Basic density is usually calculated by measuring the sample dimensions in the field, and later weighing the sample oven-dry. It is often calculated from the oven-dry density using an average value for a particular species. Appendix 6 reports the conversion factors for a number of North American species.

**Percent gray**

When DendroScan analyzes a visual sample (a sample that does not contain a calibration wedge), it records the signal as percent gray. This value is a number from 0.0 (white) to 100.0 (black) determined by linearly scaling the scanned gray values (from 0 to 255). It is related to density, but this relationship is not linear. Furthermore, it varies within the sample, depending on the color of the wood, but it rises and falls with density in such a way that a signal encoded in these values can still be used to identify ring boundaries.

**Blank years**

When trees grow in a high-stress environment, they sometimes do not form a growth ring along the entire length or circumference of the stem. As a result, there can be gaps in the annual record of some samples, particularly if they were taken near ground level. These are formally called "missing rings" but are also referred to as "blank years" in this documentation. Their existence is usually determined by cross-correlating between different samples, as the rings can be visible in some samples but not in others. DendroScan allows the user to assign as many as 3 blank years to any ring boundary in the sample. These rings are reported with a width of zero, and a density value corresponding to the average densities of the rings on either side.

**Ring boundary**

When DendroScan measures rings, it determines the ring boundary as the major inflection point in the density between the maximum latewood density of one year and the minimum earlywood density of the next (the point at which the density curve is steepest). This usually corresponds to the same location that an observer would define as the edge of a growth ring. Figure 4.1.1 illustrates this method.

Earlywood/ latewood boundary

In forestry terms, this boundary can be determined by noting when the cell walls become especially thick (often defined as that part of the ring where the cell wall width is equal to or greater than the lumina width). It also usually corresponds with changes in color and density.

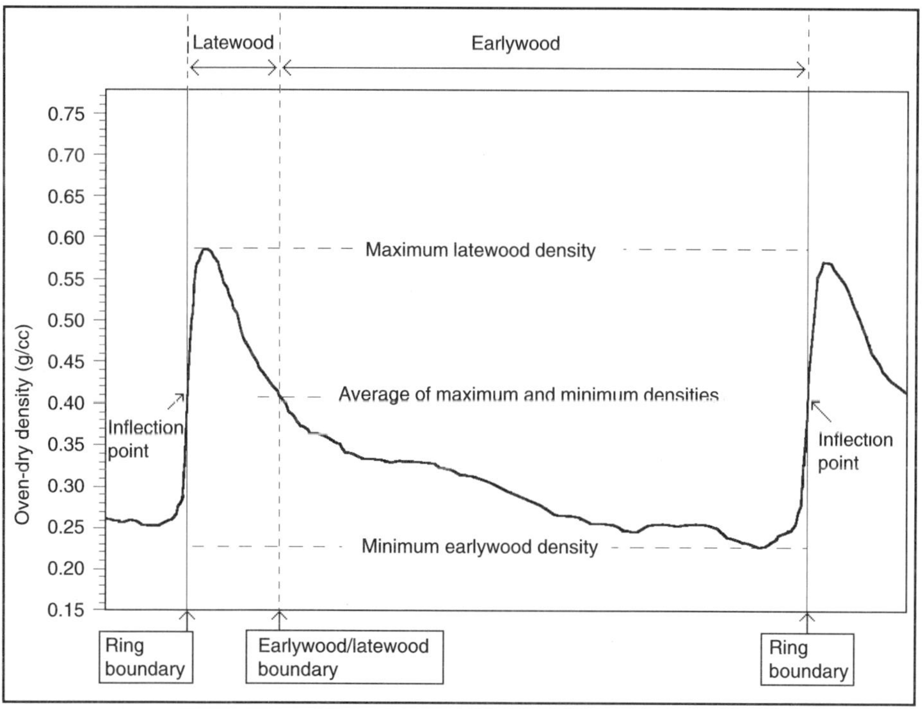


**Figure 4.1.1** **How DendroScan determines ring boundaries.** Note that the ring boundary passes through the steepest part (inflection point) of the density curve. The earlywood/latewood boundary is defined as the location between the maximum and minimum markers where the density is equal to the average of the maximum and minimum densities within a single year.

Given only density or color to go by, DendroScan cannot measure the exact boundary between the earlywood and latewood.

DendroScan instead measures an analog of this boundary using a method that works for both X-ray (where the density is known) and visual (where the density is not known) analyses. It first measures the minimum earlywood density (or gray value) and maximum latewood density (or gray value) in a given year. It then defines the boundary as the location between the earlywood and latewood where the density is equal to the average of the minimum earlywood and maximum latewood densities (*see* Fig. 4.1.1).

This results in a good analog of the biological earlywood/latewood boundary that is consistent for most rings; however, it is less accurate in the smallest rings (in rings defined by fewer than 10 pixels). Furthermore, this method will always record a value for latewood width, even in the rare years when some trees do not produce true latewood.

### 4.1.d The ISC File and Sample Header Data

When DendroScan analyzes the paths in a TIF file, it produces a linear array of density or gray-scale values complete with marks for ring boundaries. These are stored in a compact format called the ISC format. It is a compact data base that can contain up to 1000 samples, each of which can contain up to 30 000 density or gray scale values, along with ring markers. Each sample contains a header with a number of editable parameters. Some of these identifiers simply describe a sample name, while others are required for calculating reportable density values. The identifiers are as follows:

Sample number — This is a noneditable integer field (1–1000) that refers to the order of the sample within the ISC file. These numbers are used to specify which samples to include in a report.

Stand code — This is a 10-character field that can contain any alphanumeric values (all values available from the keyboard). It usually describes the forest or stand from which a sample was taken.

| | |
|---|---|
| Plot code | This is a four-character field that describes the sample plot within the stand. It can also contain alphanumeric values. |
| Tree code | A three-character field that usually contains the number of the tree within the sample plot (although letters can also be used in this field). |
| Sub code | Often replicate samples are taken, either multiple increment cores or multiple radii on a disk. This three-character field allows each replicate to be identified separately. Alternatively, it can be used to describe the disk number in a stem analysis.<br><br>The total sample code including stand, plot, tree, and sub codes (separated by dashes) is usually included in reports and graphs produced by DendroScan. |
| Last year | This is an integer field specifying the last visible year in the sample (the year immediately inside the bark). If the sample was taken late in the growing season from a live tree, this is equal to the year in which the sample was acquired. If the sample was taken early in the growing season, however, the current growth ring might not be visible, and the previous calendar year should be used. If the date of the last entry is not known, a value of 0 is recommended. This can be amended later after cross-dating. Values that are allowed can range from –9999 to 3000. |
| Sample height | This field contains the height above the root collar at which the sample was taken. In the current version of DendroScan, no decisions are made based on this value. The field exists to accommodate future additions such as a stem analysis program. |
| Species code | This four-character field is intended to contain upper-case alphabetic characters only. The program uses this abbreviation to look up Latin species names in the SPP.LST file when producing certain reports. The expected abbreviations are those specified by the ITRDB, although any four-character abbreviation referenced in the SPP.LST file can be used. These codes are also directly reported in the sample header line when reports are produced in ITRDB decadal format. |

**Distance to pith**

Often samples are taken that do not include the pith. In the case of increment cores, this is often due to sampling error. Alternatively, even in disks, the center might be missing due to core rot. This field allows the user to input the estimated distance to the pith (as measured on the image).

The distance to the pith is recorded in DendroScan as "image centimetres" (centimetres on the image regardless of scale). It refers to the estimated radial distance from the last measurable ring to the pith of the sample on the image. The program internally corrects the distance to the scale of the particular sample.

Figure 4.1.2 contains a series of concentric rings that can be used to estimate this value. When photocopied onto an overhead projector transparency, these rings can be overlain onto an X-ray positive or a sanded wood sample; the distance to the pith can be estimated by aligning the curvature of the concentric rings with the last visible rings in the sample, as shown in Figure 4.1.3.

This value is used to calculate basal area increments properly. It is also taken into account when reporting ring widths corrected to fresh conditions. As well, it is used estimate the number of rings missing from the sample (based on the average width of the last three visible rings). This allows reporting of the estimated age of the sample in the Summary Report. The allowed values in this field are from 000.00 to 999.9 cm (although in reality, the maximum distance that will allow a reasonably accurate age estimate is probably no more than 2–3 cm).

**Fresh radius**

If the fresh diameters of the tree and bark thickness are measured, the inside-bark fresh radius can be calculated and entered in the header data. This value is used when producing reports using "fresh condition". The program scales the image (allowing for rings missing near the pith) to match this fresh radius. This is most useful when a stem analysis is derived from the measurements.

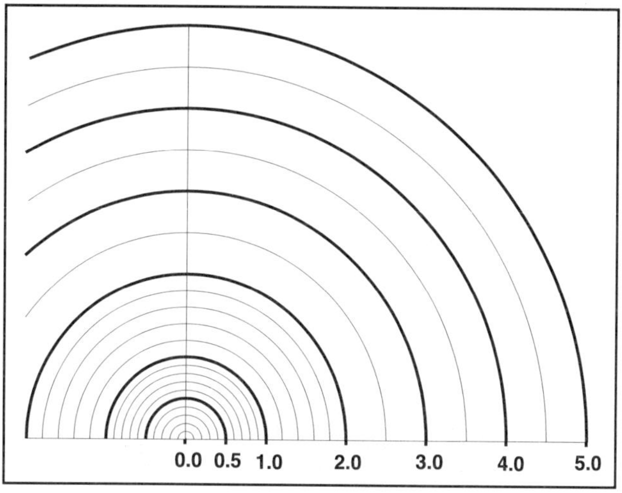


**Figure 4.1.2** **Concentric-ring tool.** Photocopy this image onto an acetate sheet. It can then be superimposed over an X-ray image to estimate the distance from the last measurable ring to the pith.

Sample thickness

The sample thickness is an important value used to correct the density values to oven-dry density. If accurate densities are required, this value must be measured with calipers on each sample. Alternatively, if the sample thin slicing is precise, a random sampling of several pieces in a picture can derive an average thickness for all the samples in that picture. The allowed values for this field are from 0.010 to 1.000 cm. This measurement can be made on fresh or dry samples, because the longitudinal dimension of wood does not change significantly with changing moisture content.

Basic density factor

The basic density factor is a multiplier that converts oven-dry density to basic density (dry mass per wet volume). This factor can be determined experimentally for each sample using volume shrinkage; however, it is usually more practical to use an average value for a given species. The values for several species are included in Appendix 6. DendroScan uses this value whenever it reports results in "fresh condition". In this way, the reported results are consistent for stem analyses and other biometric studies.

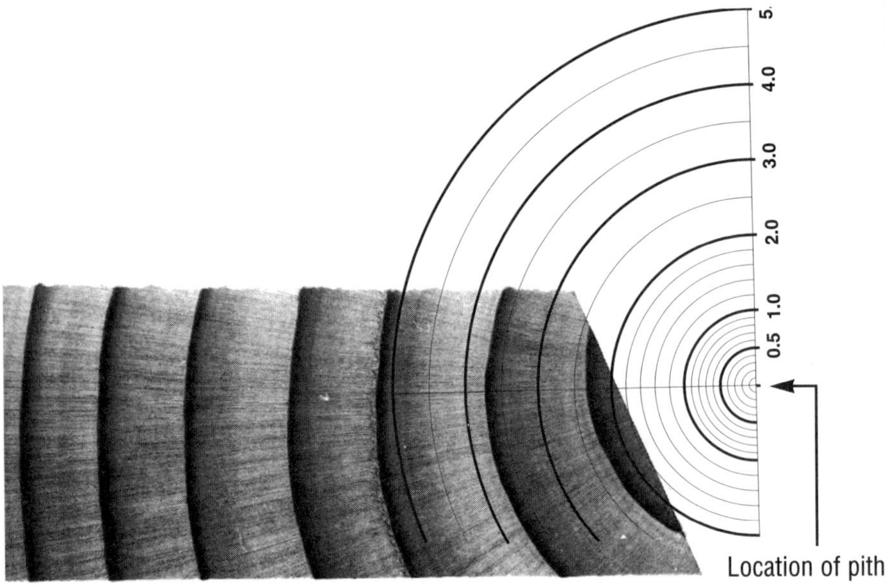


**Figure 4.1.3** **How to use the concentric-ring tool.** Superimpose the transparent tool onto the image so that the rings on the tool are parallel to the rings in the sample. The distance-to-pith is determined by the rings closest to the edge of the sample. This method assumes that tree rings near the pith are approximately circular. This assumption is usually correct enough to estimate the distance to center with an accuracy of within about 1 mm.

If the outputs are to be later converted to standard chronologies, this value might not be as important and can be ignored. The values allowed in this field are from 0.3000 to 1.0000.

## 4.2 DendroScan Graphics User Interface

DendroScan operates in VGA graphics mode. It uses a graphics user interface (GUI) that has drop-down menus and pop-up windows, similar to those found in other graphics programs such as Windows. Its operation is unique though, in that it is a DOS-based program and the GUI was custom designed. Nonetheless, the controls were designed to be similar to those used in popular commercial programs where possible. Note that this version will not support a mouse or other pointing devices, although its use of intuitive hot keys and cursor controls makes it very easy to use, even without a mouse.

### 4.2.a Menu Bar

When the program first starts up, the screen is essentially blank, except for a menu bar at the top of the screen, and a status line at the bottom. The menu bar will already be activated (Fig. 4.2.1). The menu bar consists of two main parts: the main bar, and the submenus. The menu can be activated from the main screen or the edit screen by typing the slash (</>) key or the <Escape> key. Note that on the screen, several items both on the main bar or the submenus will appear in light gray; these are menu items that are inaccessible unless an ISC file is currently open.

```
                    DendroScan VERSION 4.5
File      | Report   | Print | Utils | Sample | Edit | Help
TIF File Convert
Open ISC File
Join ISC Files

Save File
Close File
Quit Program
```

**Figure 4.2.1** **The menu bar.** This is how it appears when the DendroScan has just been started. The File submenu is open and the cursor is located at the first item of the submenu.

The menu options are as follows:

**F**ile <F1>

| | |
|---|---|
| **T**IF File Convert | Translates a TIF file and creates a sample (ISC) file (*see* Section 4.3). |
| **O**pen ISC File | Opens an ISC file for editing (*see* Section 4.4). |
| **J**oin ISC Files | Links multiple ISC files (*see* Section 4.6). |

| | |
|---|---|
| **S**ave File | Saves an ISC file while editing (*see* Section 4.4). |
| **C**lose File | Closes an ISC file after editing (*see* Section 4.4). |
| **Q**uit Program | Quits the program after editing. |
| **R**eport <F2> | *See* Section 4.7. |
| **O**utput Reports | Produces ASCII reports in several formats. |
| **P**rint <F3> | *See* Section 4.8. |
| **R**aw Density Profile | Prints graphs of the raw samples on a PostScript device. |
| **A**nnual Data | Prints graphs of various annual parameters on a PostScript device. |
| **U**tils <F4> | *See* Section 4.4. |
| **R**ecount Rings | Clears and recalculates all ring boundaries from a single sample. |
| **D**elete Current Sample | Removes a single sample from an ISC file. |
| **S**ample <F5> | *See* Section 4.4. |
| **P**rev. Sample | Shifts to a previous sample while editing. |
| **N**ext Sample | Shifts to the next sample while editing. |
| **F**irst Sample | Shifts to the first sample in an ISC file. |
| **L**ast Sample | Shifts to the last sample in an ISC file. |
| **C**ompare Samples | Allows any sample to be visually compared (ring widths) with a selected reference sample in the same ISC file (*see* Section 4.5). |
| **R**ecompare | Allows any sample to be visually compared (ring widths) with a previously selected reference sample in the same ISC file (*see* Section 4.5). |
| **E**dit <F6> | *See* Section 4.4. |
| **M**odify Sample Data | Opens spreadsheet for modifying sample header information in an ISC file. |

| | |
|---|---|
| **A**dd Blank Year | Allows the addition of a blank year (missing ring) to the current sample in an ISC file. |
| **D**elete Blank Year | Allows the deletion of a zero-width year from the current sample in an ISC file. |
| **V**iew Summary | Shows graphs of ring widths and density or latewood widths for the current sample in an ISC file. |
| **C**lose | Closes the ISC file being edited. |
| **H**elp <F7> | *See* Section 4.2.c. |
| **C**urrent Topic | Contains help on the current process being undertaken. |
| **I**SC Edit Screen | Help on keystrokes in the ISC Edit Screen. |
| **M**enu Bar | How to activate and operate the menu. |
| **T**ext Input | Help on using edit fields. |
| **H**eader Data | Defines all the data types in the ISC file header. |
| **S**pecies Names | How to select an ITRDB species abbreviation. |
| **E**rror Messages | A complete list of all error messages. |

The menu controls are as follows:

| | |
|---|---|
| **<Escape>** | If the menu is inactive, this key will activate the menu bar. If the main bar and a submenu are active, the Escape key will close the submenu. If the main bar is active and no submenus are open, the Escape key will close the menu bar. |
| **Cursor keys (→←↑↓)** | The cursor control keys allow movement both within the main menu bar and the submenus in the direction indicated by the arrows on the cursor control keys. If you move beyond the top of a submenu, the submenu will close. If you move below the bottom of the submenu, the cursor will go to the top. Moving right or left within a submenu will allow movement from one submenu to another. |
| **<ENTER>** | The program will execute the action indicated by the location of the cursor. If the cursor is in the main bar, the appropriate submenu will open. If the cursor is in a |

submenu, the action under the submenu cursor will be executed.

**Highlighted hot keys**

The letters highlighted in red in any menu item (usually the first letter) are "hot keys". If you strike any of these letters, the program immediately executes that item. Note that hot keys on the main bar are only available if the submenus are closed. Similarly, when a submenu is open, only the hot keys in that submenu are accessible.

**Function keys**

Striking the function keys will automatically open the menu and submenu appropriate to the function key pressed:

| | |
|---|---|
| **<F1>** | File |
| **<F2>** | Report |
| **<F3>** | Print |
| **<F4>** | Utilities |
| **<F5>** | Sample |
| **<F6>** | Edit |
| **<F7>** | Help |

</>

The forward slash key will activate the menu if it is closed.

### 4.2.b Status Line

The status line is a pale horizontal strip at the bottom of the screen. This line provides useful information during various operations. Sometimes it includes prompts to assist the user (for example, if the spreadsheet window is open, the status line informs the user that the <F10> key will close the spreadsheet). While data are being inputted, the status line informs the user of the current edit mode (replace, insert, or overwrite). During lengthy operations such as printing, or translating a file, the status line will show a small bar on the left side that rotates to indicate that the system is busy. This informs the user that the system is not hung, but is engaged in a lengthy operation.

### 4.2.c Help Window

Help is available throughout most of the program by simply pressing the <F7> key. In some instances, this will activate the main menu bar and allow the user to select the Help topic. In other cases, pressing this key will automatically open the Help screen to provide information on the specific operation being performed.

The Help window is basically a viewer that opens the DendroScan Help file at a location specified by the user or the context. The entire file can be viewed at any time by simply scrolling up or down with the cursor controls while the Help window is open. The Help window operational keys are as follows:

| | |
|---|---|
| **<F7>** | Opens the Help window at the current context. When viewing a DSC file, the menu bar is activated to allow a specific selection of the help desired. If the menu is inactive, this key will activate the menu bar. |
| **Cursor keys (↑, ↓, Page Up, Page Down)** | These keys allow the movement up or down within the Help file, one page at a time. |
| **<Home>** | Goes to the start of the Help document. |
| **<End>** | Goes to the end of the Help document. |
| **<ENTER>** or **<Escape>** | Either of these keys will close the Help window if it is open and return to the operation being performed prior to activating Help. |

### 4.2.d Warning/Message Windows

These windows are primarily used to display error messages, although they occasionally appear simply to signify that a lengthy or complex job was completed successfully (Fig. 4.2.2 shows a sample error message). To clear the window, simply strike any key. An explanation of warning messages appears in Appendix 4.

```
M E S S A G E

Invalid File Structure

Pick another file

< Type any key >
```

**Figure 4.2.2 A Warning window.** This particular warning is given if the user attempts to open a corrupt or non-ISC format file.

### 4.2.e Verify Windows

Verify windows will pop up in several places in the program, particularly when an action to be taken can significantly alter the contents of a file (Fig. 4.2.3).

```
              V E R I F Y
-----------------------------------------
     Save changes before closing?

        Yes                 No
```

**Figure 4.2.3** **A Verify window.** This particular verification is requested whenever an ISC file is closed.

The window contains a header that plainly states "VERIFY" and the body contains a request that requires only a Yes or No response. The cursor is usually located on the YES button in the body when the window pops up. The controls are as follows:

| | |
|---|---|
| **<ENTER>** | If the cursor is located on the YES button, then the program executes the action. If the cursor is located on the NO button, then the action will be aborted. |
| **Cursor keys** (→←↑↓) | Striking any of the cursor control keys will toggle the cursor between the YES and NO buttons. |
| **Y** | Striking the Y key will cause the action to be directly executed. |
| **N** | Striking the N key will abort the action. |
| **<Escape>** | The Escape key will abort the action. |

### 4.2.f Select Windows

Select windows are very similar in operation to the menu bar (Fig. 4.2.4 shows a sample Select window). They will pop up when a branch to several possible outcomes occurs. This window allows the user to select the particular decision.

The controls in Select windows are very similar to those for the main menu bar. They are as follows:

| | |
|---|---|
| **<ENTER>** | This key will cause the item under the cursor to be executed. |

**Cursor keys (→←↑↓)**
The cursor control keys allow movement both within the main menu bar and the submenus in the direction indicated by the arrows on the cursor control keys. If you move beyond the top of a submenu, the submenu will close. If you move below the bottom of the submenu, the cursor will go to the top. Moving right or left within a submenu will allow movement from one submenu to another.

**Highlighted hot keys**
The letters highlighted in red in any Select item (usually the first letter) are hot keys. If you strike any of these letters, the program immediately executes that item.

**<Escape>**
The Escape key will close the window without continuing the operation being performed. Under some conditions, a Verify box appears first.

S E L E C T
Select Report Type

**A**scii DendroScan
**D**igimic File
**T**ree Ring Databank RWL[a] File
**S**ample Summary
**C**ancel <escape>

[a] RWL is the file extension for the ITRDB decadal format.

**Figure 4.2.4 A Select window.** This particular window appears when creating reports.

## 4.2.g Edit Fields

Throughout the program, you will be able to enter or edit values (e.g., sample names, file names) through edit fields. When you are in one of these fields, you have the capacity to edit and change your entries in a variety of modes. The status line indicates the mode being used.

When in the replace mode (usually when you have moved your cursor to an editable field inside an Edit window or a Spreadsheet window), typing any key other than the cursor controls will erase the current entry, allowing you to create a new entry. You can get out of replace mode by simply typing the left cursor control key. This will put you into insert mode. After you have left replace mode, you usually cannot get back into it unless you move to another field and back again.

When in insert mode, typing any character will insert that character into the existing entry at the location of the cursor. If the total entry exceeds the width of the entry box, the last portion of the field will disappear to the right and will be erased as it disappears. You can toggle between insert and overwrite mode using the <INSERT> key located above the cursor controls. When in overwrite mode, typing any character will replace the character that was previously to the right of the cursor with the new character typed.

In order to minimize input errors, DendroScan limits the type of characters that can be inputted into a field, based on the type of value it expects. If you type any character that is not legal to an input field, the response will be a beep. The main fields used by DendroScan are listed in Table 4.2.1.

**Table 4.2.1 Description of edit field types**

| Field type | Characters accepted | Example 1 | Example 2 |
|---|---|---|---|
| Integer | 0123456789- | 24 | -56 |
| Real | 0123456789-. | 31.34 | -12.2 |
| Character | Any keyboard character | site # 40 | HELLO |
| Range | 0123456789-, | 1-4,7,9 | 1,5,8,11 |

Some input variables have logical limits to the values that can be accommodated by DendroScan. Attempts to input values that are too large or too small will result in error messages. For example, when specifying the thickness of a sample in centimetres, entering a value of 25.4 will result in a warning box appearing. The box informs the user that this value is outside of the acceptable range of 0.010 to 1.000 cm.

The following control keys apply to all edit fields, wherever they appear in DendroScan:

**<ENTER>** Finishes entry (the up and down cursor keys also work).

| | |
|---|---|
| **Horizontal cursor keys (→←)** | Moves cursor right or left within a string. |
| **<Escape>** | Aborts entry and replaces with original value. |
| **<Insert>** | Toggles between insert and overwrite modes. |
| **<Home>** | Moves cursor to the first character of the entry. |
| **<End>** | Moves cursor to the last character of the entry. |
| **<Delete>** | Deletes the character under the cursor. |
| **<Backspace>** | Deletes the character to the left of the cursor. |

A special edit field type (discussed in Section 4.2.1) is used for the species abbreviation. DendroScan supports the international species abbreviations currently in use by the ITRDB. In the editing of a species selection field, the use of the **<F1>** key will activate a Species Abbreviation Selection window, which allows the user to find the correct abbreviation from a list.

## 4.2.h Edit Windows

Edit windows are tables where a number of entries are required by the user for an action to be accomplished. The example shown (Fig. 4.2.5) asks for a number of parameters before printing graphs. The cursor can be moved up and down through a number of edit fields. As the cursor moves onto an item, the field will be in replace mode (if you start typing characters, the existing value will be deleted and replaced by the new entry).

The following keystrokes control the operations of an Edit window:

| | |
|---|---|
| **<ENTER>** | If on an edit field, the entry will be finished and the edit mode will change to REPLACE. If on the DONE item, the <ENTER> key will accept all entries and close the window. |
| **Horizontal cursor keys (→←)** | Moves cursor right or left within a string. If in 'REPLACE' mode, the edit mode will change to INSERT. |
| **Vertical cursor keys (↑↓)** | Moves vertically within the Edit window. If these keys are used while editing a field, the entry will be finished before |

moving away from the field. Moving above the top of the window will cause the cursor to cycle to the bottom position; conversely, moving below the bottom will cycle the cursor to the top.

**<Escape>** If in an edit field, aborts entry and replaces with original value. If on the DONE position at the bottom of the window, the Escape key causes the entire operation to be aborted.

**<Insert>** When in an edit field, toggles between insert and overwrite modes.

**<Home>** When in an edit field, moves cursor to the first character of the entry.

**<End>** When in an edit field, moves cursor to the last character of the entry.

**<Delete>** When in an edit field, deletes the character under the cursor.

**<Backspace>** When in an edit field, deletes the character to the left of the cursor.

```
D:\XCORE\TESTPINE.ISC - GRAPH PRINT DEFAULTS

Sample Numbers to Print                   1-1
Start Calendar Year                       1810
End Calendar Year                         1995
(F)resh or (O)ven Dry Condition           O
Show X Grid?                              Y
Show Y Grid?                              N
Number of SubGraphs per Graph             1
Number of SubGraphs per Page              4
Output to (F)ile or (P)rinter             P

                -- DONE --
```

**Figure 4.2.5 An Edit window.** This particular window appears when producing graphics output.

## 4.2.i File Select Windows

Whenever a file has to be selected (opening an ISC or TIF file), a File Select window will appear (Fig. 4.2.6). The File Select window lists all available files in various disks, floppies, and network drives, making it simpler for the user to input a file name. Rather than needing to remember the exact location and spelling of a file name, the user can select it from a list of existing files. The File Select window contains two subwindows: the file mask field, which controls the search method; and the file list, which lists the drives, directories, and files located. Toggling between the two subwindows is accomplished using the <F1> function key. Near the top is a directory display window that always displays the disk and directory where the search is being conducted.

The file mask is usually a mixture of legal file name characters and standard DOS wild cards. The wild cards are the asterisk (*), which represents any number of characters, and the question mark (?), which represents single characters. The file mask limits the number of files that will be displayed in the list window. For example, a mask of *.TIF would limit the search to only files with the TIF extension.

The file list contains three different selectable elements. Disk drives appear first, with each denoted by its drive letter followed by a colon and a backslash

```
      Select an ISC File
 Directory
  C:\
  USERDATA\
  File Mask (F1 to change mask)
  *.isc

 A:\
 B:\
 C:\
 ..\
 TESTPROJ\
 TESTDIR2\
SITE1.ISC
SITE2.ISC
SITE3.ISC
```

**Figure 4.2.6 A File Select window.**

(e.g., D:\). If the cursor is moved to one of these letters, the window will change to the root directory of that disk.

After the drives, the directories are listed. Directories are usually a name followed by a backslash (e.g., DS4\). As in the drive letters, hitting the <ENTER> key while on one of these directory names will switch to that directory. A special directory entry (two periods followed by a backslash [..\]) is used to move upward in the directory tree.

The remaining entries are file names. A specific file name is selected by moving the cursor to the file name and hitting the <ENTER> key. After selecting the file, a verify window will allow the user an opportunity to change the file selection. Selecting NO from this verify box will return the user to the File Select window. To exit the File Select window without selecting a file, hit the <Escape> key.

Note that DendroScan checks the file structure before attempting to load it. If the file structure does not conform to the file expected, an error message will appear and the file will not be loaded. The File Select controls are listed below.

| | |
|---|---|
| **Vertical cursor keys (↑↓)** | Moves up or down through the file list one item at a time. |
| **<Page Up> and <Page Down>** | Moves up or down through the file list one page at a time. |
| **<Home>** | Scrolls the File List window to the top of the list. Striking any key before the list has reached the top will stop the scrolling. This action is useful when searching through a long list of file names. |
| **<Control> and <Home> keys together** | Moves the file list cursor directly to the top of the list. |
| **<End>** | Scrolls the File List subwindow to the bottom of the list. Striking any key before the cursor has reached the bottom will stop the scrolling. This action is useful when searching through a long list of file names. |

**<Control> and <End> keys together** Moves the file list cursor directly to the bottom of the list.

**<ENTER>** If cursor is on a drive letter, the File List window will display the contents of the root directory of that drive. If cursor is on a directory, the File List window will change to the contents of that directory. If the cursor is on a file name, that file name will be accepted and the structure of the file will be checked before loading. The user will be asked to verify the file selection.

**<Escape>** Exits the File Select window without selecting a file.

**<F1>** Toggles between the File Mask subwindow and the File List subwindow.

**File Mask subwindow controls** The File Mask window is a character edit field. (*See* Section 4.2.g for complete controls when within this subwindow.)

## 4.2.j Spreadsheet Windows

Spreadsheet windows are similar to Edit windows except that the edit fields are arranged as a two-dimensional table. They occur in DendroScan where the user wishes to change sample header information for all of the samples in an ISC file. They allow the user to change values for a large number of header fields for multiple samples with a minimum effort. Figure 4.2.7 illustrates this particular example.

EDIT SAMPLE DATA

| Samp num | Stand code (c10) | Plot code (c4) | Tree code (c3) | Sub code (c3) | Last year (yr) | Sampl hght (m) | SPP code (c4) | Dist to pith (img cm) | Fresh radiu (cm) | Sampl thick (cm) | Dens factor |
|---|---|---|---|---|---|---|---|---|---|---|---|
| 1 | TESTPINE | 1 | 1 | A | 1995 | 1.20 | PIGL | 0.0 | 5.21 | 1.144 | 0.9000 |
| 2 | TESTPINE | 1 | 2 | A | 1995 | 1.30 | PIGL | 0.0 | 14.11 | 0.144 | 0.9000 |
| 3 | TESTPINE | 1 | 3 | A | 1995 | 1.30 | PIGL | 0.0 | 16.21 | 0.144 | 0.9000 |
| 4 | TESTPINE | 1 | 4 | A | 1995 | 1.30 | PIGL | 0.0 | 16.01 | 0.144 | 0.9000 |

**Figure 4.2.7 A Spreadsheet window.** This one allows the user to edit basic information about all the samples in an ISC file.

The Spreadsheet window operates in two main modes: the SPREADSHEET MODE, which allows the user to move between fields and samples; and the EDIT MODE, which allows editing of a given field. When in the SPREADSHEET MODE, the cursor encompasses the entirety of the field where it is located, and is colored blue. When in the EDIT MODE, all controls are as specified in Section 4.5.g, and the cursor is generally colored red. The status line helps the user to determine which mode the screen is in. If the status line displays <ENTER> to Edit <F10> to Finish, the window is in SPREADSHEET MODE. Otherwise, the status line specifies EDIT MODE.

The window will not necessarily display all of the sample headers at once. The window size changes according to the number of samples in the ISC file, but will only expand to a maximum of 20 samples visible at once. It is nonetheless possible to scroll through the remaining samples using the cursor keys.

This window occurs only in two locations in DendroScan: immediately after translating a TIF file, the allow the user to input basic header data; and from the ISC Edit screen, using the menu or function keys.

The following keystrokes control the spreadsheet operations:

**<ENTER>** If in SPREADSHEET MODE, this key activates the EDIT REPLACE MODE. If in EDIT MODE, the entry will be finished and the screen will change to SPREADSHEET MODE.

**<F10>** This key closes the window when in SPREADSHEET MODE only. All entries made are updated within the file.

**Horizontal cursor keys (→←)** If in SPREADSHEET MODE, the horizontal cursors move the cell cursor right or left from field to field within the spreadsheet. If in EDIT MODE, this moves the cursor right or left within a field. If in EDIT REPLACE mode, the edit mode will change to EDIT INSERT.

**Vertical cursor keys (↑↓)** The vertical cursors always allow movement up or down from field to field, regardless of the window mode. If the window was in EDIT MODE, the entry will be finished and the mode will change to SPREADSHEET. If the cursor is at the bottom or top of the window, and there are more samples to see in the appropriate direction, the window will scroll to view the headers of those samples.

**Page keys (Page Up, Page Down)** When in SPREADSHEET MODE, the page key allows movement up and down through the window in larger steps than the vertical cursor keys, and allows rapid movement through a large number of samples.

**<Escape>** When in EDIT MODE, aborts an entry, replacing it with the original value.

**<Insert>** When in EDIT MODE, toggles between insert and overwrite modes.

**<Home>** When in EDIT MODE, moves cursor to the first character of the entry.

**<End>** When in EDIT MODE, moves cursor to the last character of the entry.

**<Delete>** When in EDIT MODE, deletes the character under the cursor.

**<Backspace>** When in EDIT MODE, deletes the character to the left of the cursor.

### 4.2.k Species Abbreviation Selection Window

The species code is a four-character field that identifies the genus and species of tree in the file, usually the standard abbreviations specified by the ITRDB. The first pair of letters usually comprise the first two letters of the genus name, and the second pair refers to the first two letters of the species name except where this would cause duplicate abbreviations. (For example: the first two letters for *Picea* [spruce] are PC and the first two for *Pinus* [pine] are PI).

The Species Abbreviation Selection window (Fig. 4.2.8) is designed to allow the user to select the abbreviation quickly from a list published by the ITRDB. This window selectively views a previously defined list stored on disk, and allows the user to move through the list by either guessing at likely abbreviations or just browsing through the file. If the user wishes to use the ITRDB program library for later analysis or submit chronologies to the ITRDB, it is imperative that these standard names be used.

SPECIES LIST

ABGR

| | | |
|---|---|---|
| ABAM | Abies amabilis | Pacific silver fir |
| ABBA | Abies balsamea | balsam fir |
| ABCO | Abies concolor | white fir |
| ABFR | Abies fraseri | Fraser fir |
| ABGR | Abies grandis | grand fir, giant fir |
| ABLA | Abies lasiocarpa | subalpine fir |
| ABMA | Abies magnifica | California red fir |

**Figure 4.2.8 A Species List window.** Note the small inset at the upper left, which contains the abbreviation of the current selection.

The file actually used by DendroScan is a subset of this file called SPP.LST. It contains a reduced list of North American and European species. Using this subset minimizes the number of abbreviations that need to be managed at once and thus maintains speed. The user can create–edit this file using an ordinary text editor, allowing the inclusion of multinational common names (*see* instructions in Appendix 5 before attempting to create a new SPP.LST file).

This window is directly accessible whenever editing a species abbreviation anywhere in the program. Note that whenever editing an abbreviation, the status line will display <F1> TO SELECT SPECIES. Striking the <F1> key whenever this message is displayed will open this window. If the user has already started typing an abbreviation, the window will open at the appropriate location in the file.

Note that the window contains two main parts: the body (which contains a list of abbreviations, scientific names, and common names), and a small subwindow at the upper left that displays the search criteria. One line in the main body is always highlighted, and the subwindow will always display part or all of this abbreviation.

The user can move through this list using the following methods:

- Using the cursor controls will allow browsing through the file, either one item at a time using the arrow keys, or one screen load at a time using the page keys. When these keys are used, the abbreviation in the subwindow always reads the same as indicated in the highlighted line in the body.

- The user can start typing a guess using the alphanumeric keys. The body will shift to that part of the list that most-closely approximates the spelling of the abbreviation typed. Once four letters have been typed, inputting more letters clears the abbreviation previously input and replaces it with the new letters typed.

The following keys control the species abbreviation selection screen:

**<F1>** If editing a species abbreviation, this key will open the Species Abbreviation Selection window. If no abbreviation had previously been typed, the window will open at the start of the list. If some letters had already been typed, the window will open at that part of the list that most-closely approximates the spelling of the letters already typed.

**Cursor keys (↑↓)** These keys move the main cursor up or down through the list, one item at a time. Any partially typed abbreviation displayed in the subwindow is immediately replaced by the entire abbreviation of the new item highlighted.

**Page keys (Page Up, Page Down)** These keys move the main cursor up or down through the list, one screen load at a time. Any partially typed abbreviation displayed in the subwindow is immediately replaced by the entire abbreviation of the new item highlighted.

**<Home> and <End> keys** The <Home> and <End> keys move the screen to the start or end of the list, respectively. Any partially typed abbreviation displayed in the subwindow is immediately replaced by the entire abbreviation of the new item highlighted.

**Alphabet keys (<A> to <Z>)** As these keys are struck, an abbreviation string will appear in the subwindow, and the body will spontaneously shift to that part of the list that most-closely approximates the spelling of the abbreviation currently displayed. Once four letters have been typed, inputting more letters clears the abbreviation previously input and replaces it with the new letters typed.

| | |
|---|---|
| **<ENTER> or <Escape>** | Either of these keys will close this window and return to the edit field where the user started, replacing the field with the new abbreviation selected from the list. |

## 4.3 Translating a TIF Image File

Translating a TIF image file is done with the main menu bar (<**F**>ile -<**T**>IF File Convert). This is used to read the files that were created by the scanner software. This section will translate the images into density traces (or gray-scale traces in the case of visual images) and create a compressed image file (ISC format) that will be used for later processing. This section also allows you to link samples that were too long to fit onto a page, and automatically measures and counts all the rings for each sample. At the end of the translation, the user will be allowed to edit the sample header information, and enter directly into an ISC edit session.

### 4.3.a The TIF Translation Screen

A File Selection window will initially open. When a valid TIF file is selected, the TIF translation screen will then open. This screen is essentially a window on the file, showing the progress made by DendroScan and allowing the user to make selections. It is a visual aide that provides necessary feedback for the following:

- a final examination of image quality and placement of samples;
- correcting some errors related to misplaced reference lines; and
- linking several segments of samples that are too large to fit on the scanner in one piece.

In this section of the program, DendroScan displays scan lines as it is processing the TIF file. The samples are displayed in 12 shades of brown, plus white and black to display extremes. Blue lines show the lines initially drawn during the scanning process. Lines actually used for processing are redrawn in red as the gray-scale/density-averaging occurs.

When DendroScan reaches the end of the first sample, it will ask the user if the first sample is a calibration wedge. If the user replies Yes, the assumption is made that the image is an X-ray sample. An Edit window will then request the wedge number. The initial default should be 1 unless more than one wedge has been calibrated by the user (*see* Appendix 5). After the wedge has been identified, a dotted red line will appear across the image of the wedge on the screen. The status line will announce that the image is being calibrated.

The file will then be scanned line-by-line and displayed on the screen, with path lines drawn in blue. As each sample is found, it is announced on the status line. The blue lines then turn red, proceeding from left to right. The red line is the path that the program is following to analyze the image. Notice that elevator lines stay blue, as they are not used as path designators. If an elevator turns red, or if the red line for some reason strays from the blue line, the program is having problems following the path designated by the user.

Note: If the elevator lines turn red, or if a red line appears in any other area but on top of a blue line, do not ignore it. Either of these conditions is an indication that the rules for drawing lines have been broken (*see* Section 3.3). The processing can be stopped at any time by typing any key. This will trigger a Verify box that asks if the user wishes to continue processing (Yes) or abort the operation (No).

### 4.3.b Linking Samples

After the entire image is displayed, numbered blue boxes will appear around each sample, and a verify window will ask if any samples need to be linked. If a sample was originally broken into two or three main pieces (because it was too long to fit on the X-ray as one piece), it can now be reconnected. Alternatively, sometimes the program sees one sample as two pieces (due to a lack of use of elevator lines) and they need to be joined as one. This section of the TIF translation will allow the user to join the pieces appropriately.

Answering Yes generates an Edit window requesting the number of the prefix and suffix sample numbers. The resulting samples entered will be linked in the order indicated. The following rules apply to linking samples:

- Samples can only be linked if they are consecutive in the image (e.g., 1 and 2). Attempts to enter nonconsecutive samples will result in a warning, and the action will not be taken.
- Samples will be linked in the order entered: the prefix sample first, and suffix sample afterwards. Be careful of the order in which the samples are entered.

Note: After each link of two samples, the samples will be renumbered. Take note of the new numbers as displayed on the screen before linking the next pair of samples.

Only two samples can be linked at a time; however, the user might wish to link several samples. This can still be accommodated. If, for example, the user wishes to link samples 1, 2, and 3 (in that order), samples 1 and 2 should be linked first. After the first link, the samples are renumbered; the first two samples will now be called sample 1 and the former sample 3 will now be called sample 2. The second link should therefore be samples 1 and 2 once more to complete the link. When the samples are renumbered again, the three original samples will be linked in consecutive order, and will thereafter be treated as sample one.

When linking is complete, answer No to the next request for linking.

### 4.3.c Edit Default Data

After linking operations are complete, an edit window will appear entitled EDIT DEFAULT DATA. This window displays the default data that will be entered into the headers of every sample. This is intended to prevent unnecessary repetitive entry of the same number into every sample. (For example, the basic density factor or sample thickness). The editable defaults are listed below. Details on entering and/or editing values are given in Section 4.2.h.

The list of editable default fields is as follows:

- Stand code (10-character field)
- Plot code (four-character field)
- Tree code (three-character field)
- Sub code (three-character field)
- Last year (four-digit numeric field)
- Species code (four-character alphabetic field)
- Sample thickness (real number field, cm)
- Basic density factor (real number field, no units)

If left unchanged, the displayed last year will have a value equal to the year on the computer's clock minus one. Species code should be adjusted or its value will default to a nonsensical AAAA. The sample thickness value that first appears is that indicated in the DENDSCAN.INI file. The basic density factor appears as 0.900. It should be changed to a value more appropriate to the species. More details on the appropriate entries for these fields is provided in Section 4.1.d.

### 4.3.d Edit Sample Header Data

Next, a Spreadsheet window entitled EDIT SAMPLE DATA will appear. The user has the opportunity here to enter detailed header information for each sample. The list of fields that can be edited from this spreadsheet are listed below.

- Stand code (10-character field)
- Plot code (four-character field)
- Tree code (three-character field)
- Sub code (three-character field)
- Last year (four-digit numeric field)
- Sample height (real number field, m)
- Species code (four-character alphabetic field)
- Distance to pith (real number field, image cm)
- Fresh radius (real number field, cm)
- Sample thickness (real number field, cm)
- Basic density factor (real number field, no units)

Note that in the current version of DendroScan, the sample height is not used and so is an optional field. To derive accurate measures of density, the sample thickness should be measured accurately for each sample and recorded here. Fresh radius normally defaults to the total length of the sample that DendroScan measured during analysis. Details on editing the fields are given in Section 4.2.k and details on the most appropriate contents of these fields are discussed in Section 4.1.d.

After completing any editing, the user should strike the <F10> key to close this window. DendroScan will then create a temporary ISC file, mark all the ring boundaries on the samples, and open the ISC Edit screen. Remember to save your file after completion.

## 4.4 Editing an ISC Data File

ISC files are essentially long lists of density values, along with markers to define ring boundaries, and headers to define basic sample information. There can be many thousands of density values for a given sample. DendroScan provides a graphical, user-friendly interface to supervise and overriding the program's decisions in marking ring boundaries, and allows sample comparisons and production of graphs and reports.

**Tips for editing samples**

- Go through each sample and correct any obvious ring problems. It will be useful to have the picture handy as well as the negative and a microfiche reader (to view the negative). Use the sample summary often to identify suspicious rings.
- Then cross-correlate samples and correct any discrepancies among samples.
- Go back and forth between samples looking at the raw data. Often there are marker years, which are years with unusual features.
- If a sample does not correlate, examine the original; sometimes part of the sample became lost during processing. At other times, a piece of a sample was X-rayed out of order or was backward on the image. If it is not possible to determine the problem, delete the sample from the data set (after making notes to that effect so that the reason for deletion is clear).

### 4.4.a The Ring Edit Screen

When an ISC file is opened (or a TIF file has just been converted), the screen opens to the Sample Edit screen. This screen is used to verify and edit ring boundaries. Figure 4.4.1 illustrates this screen and its components.

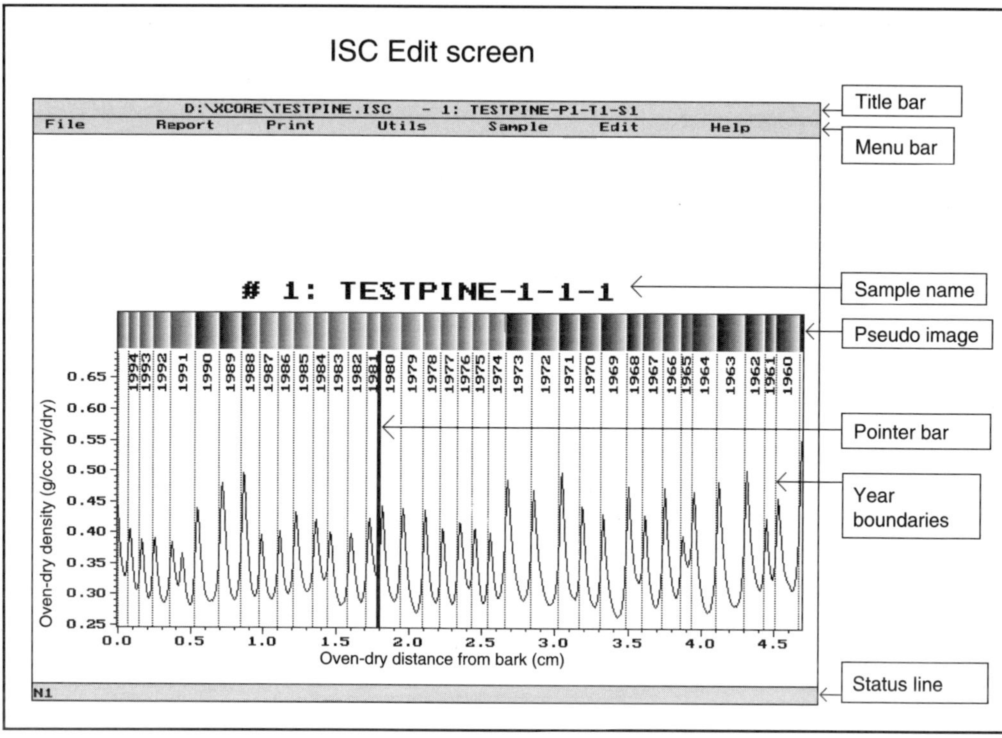


**Figure 4.4.1** **Illustration of the ISC Edit screen.** The inset boxes describe the basic parts of the screen.

The graph is a simple measure of density (or gray intensity) plotted against distance from the bark. The density is drawn in solid black lines, and ring boundary markers as dashed vertical blue lines. Numbers written vertically inside the graph indicate the calendar year represented by the space between a given set of boundaries. The entire graph is usually not seen on the screen at once. Imagine the graph to be a long strip up to several metres long, and the computer screen is a window looking at a portion of the graph. The strip can then be slid back and forth to see different parts of the graph through the window.

At the top of the graph is a multishaded bar. This is a pseudo-image of the sample that helps orient the user within the sample. It is generated using a combination of the density curve and a statistical representation of the color variance within the TIF file from which the density was derived. As such, it provides not only information on density, but on texture and continuity as well. As with other screen images in this program, the shade bar is composed of only 13 colors. The color separation is automatically adjusted by DendroScan, however, using the density variance in a particular sample and ensuring good contrast and visibility of discontinuities. Sometimes, the detail visible here is superior to the original scanned image, due largely to this contrast adjustment.

The horizontal axis is fixed at 550 pixels wide, although the scale is always indicated in centimetres from the bark. The vertical axis is automatically scaled to encompass all of the density variations within the entire sample.

### 4.4.b Editing Within a Sample

The vertical red line defines the pointer that can be moved throughout the graph. Moving this pointer back and forth through the graph allows the user to observe the locations of selected annual ring boundaries and make corrections as necessary. To remove a ring mark, move the cursor to the mark and hit the space bar. To add a ring mark, move to the appropriate location and hit the space bar again. In high-quality samples, very few overrides are required, although it is highly recommended that the entire sample be examined.

Note that as the pointer is moved, letters and numbers might appear at the left side of the status line. These codes designate peaks and valleys in the density signal, as well as the ring boundaries. They can also be useful information for the user. N1 denotes a ring boundary. If the number is higher than 1, then missing rings (blank years) have been marked at that boundary. P indicates a peak marker. If it is in upper-case, then it designates the location of the maximum latewood density for the year in which it occurs. V indicates a valley. If it is in upper-case, then it indicates the location of the minimum earlywood density for a given year.

The following summarizes the keys and menu items used for moving the pointer bar and marking/unmarking ring boundaries:

**Horizontal cursor keys (→←)** The horizontal cursor keys allow rapid movement of the pointer bar from ring boundary to ring boundary. This is useful for rapid scanning of the file.

**Vertical cursor keys (↑↓)** The horizontal cursor keys also allow rapid movement of the pointer bar, but from inflection point to inflection point. Inflection points are areas in the graph where the density is rising steeply. These might be obvious, as in ring boundaries, or less-obvious when hidden among the noise within a ring. When trying to insert a new ring boundary, moving with the vertical cursors moves to locations where ring boundaries are likely to occur.

**Control plus the horizontal cursor keys (Ctrl →, Ctrl ←)**
By holding the control key and striking one of the horizontal cursor keys simultaneously, the pointer bar can be moved slowly, one pixel at a time. This is useful when trying to locate the best spot to mark a less-obvious ring boundary.

**<Home> and <End> keys**
These keys move the pointer bar directly to the bark edge (home) or to the pith (end).

**Page keys (Page Up, Page Down)**
The page keys allow rapid movement across different sections of a sample, as they move the pointer bar 250 pixels at a time (one half-screen). The Page Up key moves the pointer right and the Page Down key moves the pointer to the left.

**Space bar**
Striking the space bar performs one of three tasks. If the pointer is at a current boundary, it causes that boundary to be removed. If the boundary contains missing rings (indicated on the status line as N followed by a number 2 to 4), one of these rings will be removed. Lastly, if the pointer is located at any other location, a boundary marker will be inserted. (Note that DendroScan will not allow a boundary marker to be inserted at a pixel location designated as a valley.)

**Missing rings (blank years)**
Missing rings can be inserted using the menu: <**E**>dit, <**A**>dd blank year. They can be removed using <**E**>dit, <**D**>elete blank year. Note that blank years can only be added at an existing ring boundary because by definition, they must exist between existing rings. A maximum of 3 blank years are allowed at any single ring boundary.

### 4.4.c Switching Among Samples in an ISC File

ISC files can contain many samples (up to 1000), and it is useful to move among samples to compare various parameters and the location of particular ring markers. Often the user will want to edit several samples at once to cross-date them. Switching among samples can be accomplished within the Edit screen using keystrokes or using the menus. The pointer bar is associated with a particular sample. Thus, when switching back and forth between samples, the pointer bars for each sample will not move unless explicitly moved by the user. This makes it easy to compare several samples. The following lists the methods of moving rapidly from sample to sample within an ISC file.

**Control plus the page keys** By striking the Control and one of the page keys, it is possible to switch to an adjacent sample. Ctrl-Page Up switches to the previous sample, and Ctrl-Page Down switches to the next sample. This can also be accomplished through the menu using <**S**>ample, <**P**>rev Sample or <**S**>ample, <**N**>ext Sample.

**Control plus the Home and End keys** Ctrl-Home causes switching to the first sample in the file. Similarly, Ctrl-End switches to the last sample in the file. These can also be accomplished through the menu using <**S**>ample, <**F**>irst Sample or <**S**>ample, <**L**>ast Sample.

Note that it is also possible to move among samples while viewing summaries (Section 4.4.d) or cross-correlating (Section 4.4.e).

### 4.4.d Viewing a Sample Summary

It is often useful to view a summary of the ring widths and densities of a sample on an annual basis. This can be accomplished for any sample using either the following menu or hot keys:

<**E**>dit, <**S**>how Summary — Menu—Open the Default Summary window.

<**F8**> — Open the Default Summary window.

<**Ctrl**><**F8**> — Choose between viewing latewood or density.

Figure 4.4.2 illustrates the Summary screen. This particular example shows oven-dry density on the upper graph, and ring width on the lower graph. In the case of a visually analyzed sample where density is not available, the default screen will show relative latewood width as the top graph. This parameter can be viewed for X-ray samples as well.

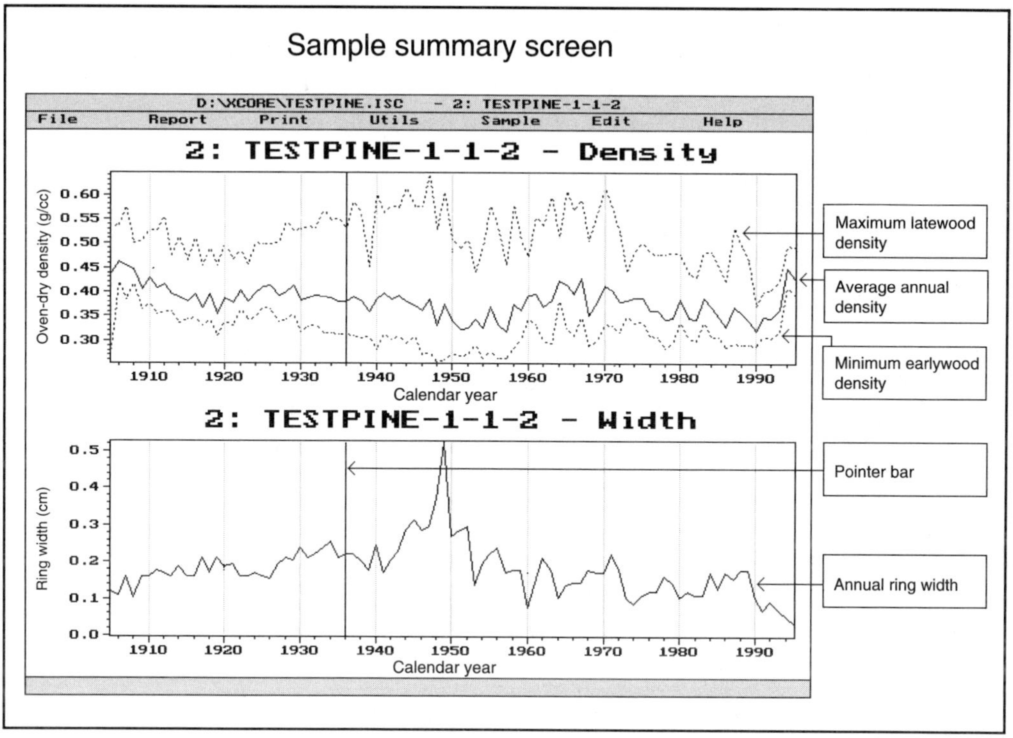


**Figure 4.4.2** **Illustration of the Sample Summary screen.** An examination of abnormalities in this screen will assist the user in identifying mismarked ring boundaries.

Note that the direction of the graph appears backwards from the main Edit screen. In the Summary screen, years advance chronologically from left to right, whereas they appear to advance in the reverse direction in the main Edit screen. In the main Edit screen, the horizontal axis is distance from the bark, because the bark is often the best reference point for measurement. This results in the most-recent year being displayed on the left. The authors chose to display the Summary screen with lowest number to the left as in most conventional graphs, because the horizontal axis of this screen shows calendar years. The apparent chronological reversal between these screens can be slightly disorienting at first.

The entire file is always displayed, and the vertical and horizontal axes are auto-scaled to fit on the screen. The pointer bar in this view is connected to the pointer in the main Edit screen. Moving the bar from year to year will move the Edit screen pointer to the starting ring boundary of the same year on the main Edit screen. Thus, if a suspicious density value or ring width appears on the Summary screen, moving the pointer bar to this year, and exiting to the main Edit screen, will allow the user to identify anomalies quickly and then examine them in more detail.

It is also possible to move among samples from the Summary screen using the vertical cursors. These and other operations are controlled using the following keys:

**Horizontal cursor keys (→←)** The horizontal cursor keys move the pointer bar one year at a time within the Summary screen. Note that this also moves the pointer bar on the main Edit screen to the same location.

**Vertical cursor keys (↑↓)** The vertical cursor keys switch to the previous (up arrow) or next (down arrow) sample. This switching among samples is also linked to the main Edit screen.

**<F8> function key** This key produces a Select window that allows the user to switch the upper graph display between oven-dry density and relative latewood width. Note that if the sample being analyzed is not an X-ray, only relative latewood width can be displayed.

**Any other key** Striking any other key except the above will close the Summary screen and re-open the main ISC Edit screen. Note that if the pointer bar was moved within the Summary screen, it will also have moved within the main Edit screen.

### 4.4.e Cross-correlating Samples

After each sample has been checked and ring boundaries adjusted, comparing the annual signature of ring widths is indispensable as a final tool for locating missing years, false rings, and other anomalies. A comparison screen can be activated from the main Edit screen using the following menu or hot keys. (Note that the first time this screen is accessed in any particular edit session, any of the above keys will first request that a reference sample be specified before the screen is opened.)

| | |
|---|---|
| <**S** >Sample, <**C**>Compare Samples | Menu—Select new reference sample before opening Compare Sample screen. |
| <**Ctrl**><**F9**> | Hot-key access to the above item. |
| <**S**>Sample, <**R**>Recompare Samples | Menu—Open the Compare Sample screen using the default reference sample. |
| <**F9**> | Hot-key access to the above item. |

Figure 4.4.3 illustrates the Compare Sample screen. It resembles the Sample Summary screen in that two graphs are displayed; however, the upper graph is shaded, and both graphs display ring width. The upper graph displays a reference sample. The lower graph displays the current sample. The current sample is the sample that was visible on the main Edit screen prior to accessing the Compare screen. The reference sample should be selected carefully by the user; it should represent the sample in which the user has the greatest confidence in the selected ring boundaries.

Moving the pointer bar from year to year can help determine if there are any differences between the two graphs where an anomaly may exist. The easiest way to discover the anomaly is to move the pointer to where the graphs appear to go out of phase, and then drop back to the main Edit screen. Similar to the Sample Summary screen, the pointer bar in the current sample is connected to the pointer in the main Edit screen. Thus, the above procedure will quickly identify where mismarked ring boundaries occur. As in the Summary screen, the horizontal axis in the Compare Sample screen appears reversed to the main Edit screen.

As in the Sample Summary screen, it is possible to move among samples from within this screen. The controls are also similar to the summary screen controls and are as follows:

| | |
|---|---|
| **Horizontal cursor keys** (→←) | The horizontal cursor keys move the pointer bar one year at a time within the Summary screen. Note that the pointer bar is connected to the pointer bar on the current sample in the main ISC Edit screen. The reference sample pointer will always remain fixed. |

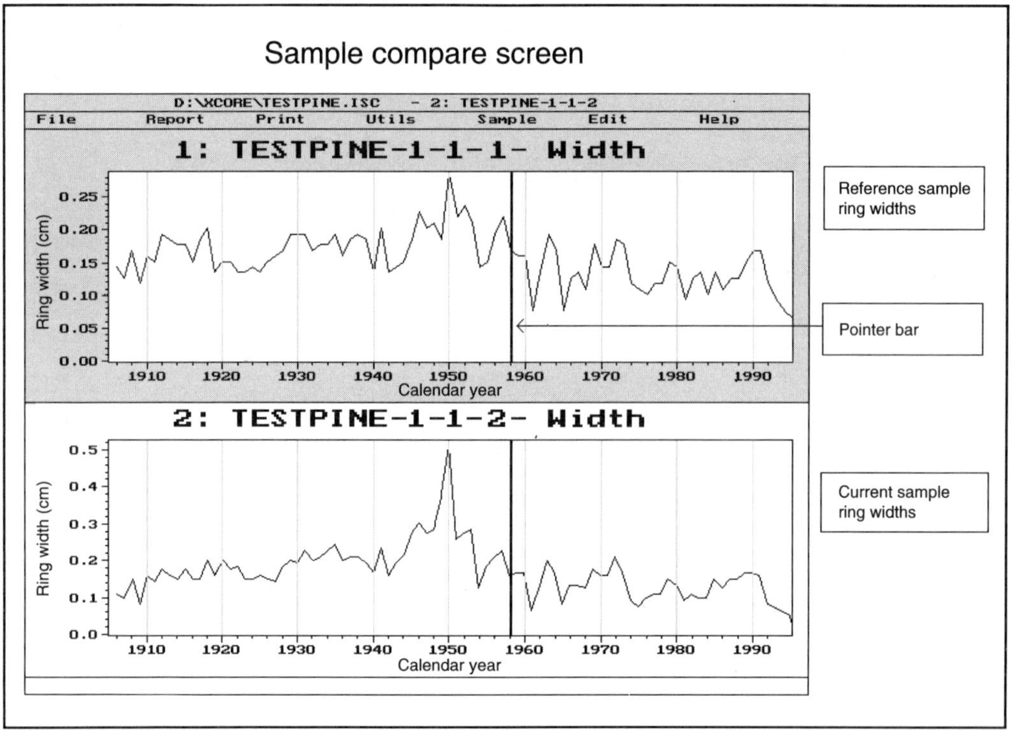


**Figure 4.4.3** **Illustration of the Sample Compare screen.** This screen allows the user to compare the ring-width response in various samples within an ISC file. This is particularly useful in discovering missing or otherwise mismarked rings.

**Vertical cursor keys (↑↓)** The vertical cursor keys switch the current sample to the previous (up arrow) or next (down arrow) sample.

**<F9> function key** This key produces an Edit window that allows the user to select a different reference sample.

**Any other key** Striking any other key except the above will close the Compare screen and re-open the main ISC Edit screen. Note that if the pointer bar was moved within the Compare screen, it will also have moved within the main Edit screen.

### 4.4.f Modifying Header Data

When an ISC file is first generated from a TIF file, the user has the opportunity to modify header information for each sample using a Spreadsheet window. This window can be re-opened at any time from the ISC Edit screen using the <**F10**> hot key or the menu (<**E**>dit, <**M**>odify Sample Header). Details on using this screen are included in Section 4.1.d.

### 4.4.g Other Editing Operations

There are a few other operations that can be performed from the Edit screen. Some should be conducted with care, as they might involve the loss of data; others (such as saving the file) should be done often.

**Reset ring markers (recount rings)**

If the user becomes frustrated with marking and unmarking rings and wishes them to be reset back to the original markers selected by DendroScan, this can be accomplished by using the menu <**U**>tils, <**R**>ecount Rings.

**Deleting a sample from an ISC file**

Occasionally, a sample needs to be deleted due to poor X-ray imaging or other problems. It is possible to delete an entire sample from an ISC file by first moving to the sample within the Edit screen, and then using the menu item <**U**>tils, <**D**>elete Current Sample.

**Recovering from a deletion or ring marker resetting**

Both of these operations should be well-thought-out before executing. If a sample has been deleted, or the ring markers reset by mistake, it is possible to recover the sample because DendroScan always works with a temporary file. To recover a sample, simply close the ISC file without saving it, and then re-open the same file. The sample will be recovered, but the editing session that was not saved will be lost. This task can be accomplished from the menu by first executing <**E**>dit, <**C**>lose. When the Verify window asks if you wish to save the file before closing, press the <**N**>key. Then re-open the file using <**F**>ile, <**O**>pen ISC File menu option.

**Saving the file while editing**

The fewest problems will occur if a file is saved often during editing. To perform this task, use the <**F**>ile, <**S**>ave menu option.

## 4.5 Producing Reports

Reports can be produced in a number of ASCII formats using the main menu (<**R**>eport, <**O**>utput Reports). A Selection window will appear, allowing a choice of several types of reports. Three report types are unique to DendroScan. One report format mimics the Output from a Digimic ring measurement system. Reports compatible with the ITRDB program library are also available. The user is also given the option to report values based on oven-dry or fresh conditions.

### 4.5.a Report Parameters

With all report types, an Edit window allows the user to specify which samples to include in the report, and whether to report using oven-dry, or fresh conditions. Later, the user will be allowed to specify the name of the report file.

**Sample numbers to print**

This edit field will appear as an option in an Edit window just after the report type has been selected. It expects a range of sample numbers. The default response will be all the samples in the file.

**(F)resh or (O)ven-dry conditions**

This edit field appears in the same Edit window as the sample numbers. If O is inputted into this field, ring widths will be measured in the raw condition, as determined by the scan. Density will be measured as oven-dry. All parameters based on width and density will use dry conditions. If F is inputted into this field, widths will be scaled according to the fresh sample diameter input in the sample headers. Density will be adjusted to basic density using the basic density factor specified in the header.

The best option to select depends on the intended purpose. If dimensionless tree-ring indexes are to be developed, the option picked doesn't matter much, because only relative values are required. For stem analysis, however, fresh conditions are more appropriate, because the outputs will most-closely resemble the stem measurements made in the field, and biomass will be more accurately portrayed.

**Report file name**

This edit field appears in a separate Edit window. The default file name usually depends on the report type selected. In most of the reports, the default file name is derived from the name of the ISC file, with a change in the three-character extension. In the case of RWL[3] reports, the first three letters of the ISC file are used, followed by a letter code signifying the data contained, and the extension RWL. In all cases, the default directory is usually the same directory in which the ISC file resides.

### 4.5.b ASCII DendroScan Report

The ASCII DendroScan report is a standard ASCII columnar file. If the original samples are from a visual analysis, the report will contain three columns: calendar year, ring width (in cm), and relative latewood width. If the ISC file contains X-ray samples, then the report will have an additional three columns, including maximum latewood density, minimum earlywood density, and average annual density (all in g/cc). Each sample will have a title, and each column will have a header.

This file can easily be imported to a spreadsheet such as Lotus, Excel, or Quattro for further data manipulation.

### 4.5.c Digimic Format Report

The Digimic format contains only ring widths. Note that the sample names have to be carefully identified, as this format assumes that only numeric sample names exist (*see* Appendix 2).

### 4.5.d Tree Ring Data Bank RWL Format Report

This format is called DECADAL FORMAT by the ITRDB, and is the main input format for calculating standard ring indexes, for correlation to climate and environmental parameters. Each line of the file includes 10 years worth of data, and each file can only contain one type of data (although it can contain many samples). Note that the sample-naming structure required for this format is restricted to six digits. To overcome this incompatibility, the DendroScan sample name appears in a comment line at the start of the sample, along with the ITRDB species code and a description of the data. The actual sample name appearing at the left side of each line is composed using the first three digits of the ISC file name, followed by a three-digit sample number padded with zeros (Appendix 2). This format allows for a versatile output, and thus allows the user the most choice in parameters to report. Table 4.4.1 summarizes the data that can be outputted to RWL format.

---

3 RWL is the file extension for the ITRDB decadal format.

**Table 4.4.1 Summary of data types in International Tree-Ring Data Bank RWL format reports**

| Data group | Data type | File ending | Comments |
|---|---|---|---|
| WIDTH | Total ring width | _RW.RWL | Ring width (cm) |
| WIDTH | Earlywood width | _EW.RWL | Earlywood width (cm) |
| WIDTH | Latewood width | _LW.RWL | Latewood width (cm) |
| WIDTH | Relative earlywood | _REW.RWL | Earlywood width/ring width (unitless) |
| WIDTH | Relative latewood | _RLW.RWL | Latewood width/ring width (unitless) |
| DENSITY | Ring average | _RD.RWL | Average density in the ring (g/cc) |
| DENSITY | Earlywood average | _ED.RWL | Average density in the earlywood (g/cc) |
| DENSITY | Latewood average | _LD.RWL | Average density in the latewood (g/cc) |
| DENSITY | Minimum earlywood | _MND.RWL | Minimum density in the earlywood (g/cc) |
| DENSITY | Maximum latewood | _MXD.RWL | Maximum density in the latewood (g/cc) |
| OTHER | Basal area increment | _BAI.RWL | Basal area increment ($cm^2$) |
| OTHER | Basal mass increment | _BMI.RWL | Basal area increment × average density ($g/cm^2/cm^3$) |
| OTHER | Ring mass | _RM.RWL | Ring width × average density ($g/cm/cm^3$) |

## 4.5.e Raw Profile

The raw profile report is essentially a dump of the density profile for one or more samples. It is of limited use in later analysis. These reports should be produced only if necessary because they can be extremely large.

## 4.5.f DendroScan Summary Report

The DendroScan summary report option produces a file containing summary values for each sample. The following information is included:

- Sample number
- Sample ID (including stand, plot, tree and sub codes)
- ITRDB sample code (the sample identifier that would be outputted to RWL format files)
- ITRDB species code
- First year
- Last year
- Total rings in the sample
- Raw dry radius (cm)

- Pixel resolution (mm)
- Distance missed to pith (in oven-dry cm)
- Adjusted dry radius (sum of raw dry radius and distance missed to pith)
- Estimated age (adjusted based on distance missed to pith)
- Fresh radius (as input in the header)
- X-ray thickness (thickness of the X-ray sample in cm)
- Average oven dry density of the sample (g/cc)
- Basic density factor
- Average basic density of the sample (g/cc)

The information in this report is useful for checking on analysis quality, determining ages of trees from increment cores, and cross-relating the sample names used in DendroScan with those reported in files submitted to the ITRDB.

## 4.6 Printing Graphs

Graphs can be printed directly to a PostScript device from DendroScan. Such devices include a number of laser jet and ink jet printers, as well as several plotters. PostScript is a machine-independent, vector-based page description language. The authors chose this language to make printing possible on a number of platforms without the need for employing hundreds of machine-dependent drivers. If the user has no access to a PostScript device, outputs can be directed to a file for later printing.

Graphs are produced using the main menu (<**P**>rint). Two main options are available: <**D**>ensity Profile and <**A**>nnual Data. The first option produces a raw-data graph similar to what is displayed in the sample edit screen. The second option opens a New Selection window with options for producing a number of annual data graphs. Appendix 2 contains samples of the different graphs produced.

### 4.6.a Density Profile Graphs

The density profile graph is an output of the entire profile of a particular sample. The vertical axis represents density (or gray-scale) and the horizontal axis represents distance from the bark in cm. Given that this graph is complex and contains a large number of points, the user has the option of breaking it up into sections so that individual ring details can be easily seen. The vertical axis is automatically scaled so that all density values are visible in the graph. The user controls the horizontal axis scaling by specifying the number of cm to be displayed in each graph. Ring boundaries are displayed as vertical bars, and every fifth year is labeled diagonally at the top.

| | |
|---|---|
| **Sample numbers to print** | This field expects the user to input the specific samples to be sent to the printer, and expects a range of file sample numbers. The default response will be all the samples in the file. |
| **(F)resh or (O)ven-dry conditions** | If O is input into this field, ring widths will be measured in the raw condition, as determined by the scan. Density will be measured as oven-dry. All parameters based on width and density will use dry conditions. If F is input into this field, widths will be scaled according to the fresh-sample diameter input in the sample headers. Density will be adjusted to basic density using the basic density factor specified in the header. The default value is oven-dried (O). |
| **Show Y grid (Y/N)** | This asks the user if horizontal grid lines should be printed across the graph at each tick on the vertical axis. The default is No. Note that the user is not given this option for the X axis; this is because vertical grid lines are automatically printed to represent annual ring boundaries. |
| **Subgraph width (cm)** | This field allows the user to specify the scaling of the horizontal axis, not the physical width of the graph on the page (which is fixed). It is an integer field, so that a whole number of cm should be used to define the width of a given graph. If, for example, a sample is 17 cm long, inputting 4 cm would cause the graph to be split into five sections, each 4 cm wide. Each sample could have a different number of subgraphs because different samples are different lengths. The default value for this field will be different for each file, and is determined by dividing the width of the widest sample in the file by four, and rounding up to the nearest whole cm. |
| **Subgraphs per page** | This integer field allows several subgraphs to be stacked on a page. The height of each graph is adjusted |

accordingly. Thus, this also gives the user some measure of control on the physical size of the printed graph. The default is 4.

**Output to file or printer** This selection allows the user to specify whether to send the file directly to the printer or save it as a PostScript file for later printing. The default option is P (send directly to printer). If the F (send to a file) option is used, a separate Edit window will request the name of the output file. This file can be printed later when a PostScript device is available, or a translator program such as GhostScript can be used to convert the file for printing to something other than a PostScript device.

### 4.6.b Annual Data Graphs

There are a number of annual graphs that can be produced. These include the following:

**Summary of ring width and density** This is a summary of three graphs stacked on a page. The first graph is ring width (cm), the second graph is density (maximum latewood, minimum summerwood, and annual average in g/cc), and the third graph is relative latewood width (unitless).

**Width** Ring width is displayed (cm).

**Density** This graph displays a mixture of maximum latewood density, minimum summerwood density, and annual average density on the same graph.

**Relative latewood width** This displays the relative width of the latewood (latewood width divided by ring width).

**Ring mass** This displays the product of ring width and average ring density. The profile is similar to ring width.

**Annual density profile** This is a specialized graph that produces a profile similar to the profile described in Section 4.6.a. It differs in that the profile for each year is normalized to the same width, and the horizontal axis is calendar years rather than distance from the bark. This graph is particularly useful when attempting to compare several samples using specific ring details, because intra-year events such as frost rings will usually occur in the same relative position in a growing season.

A number of parameters can be selected for annual data graphs. They differ slightly from the parameters mentioned in Section 4.6.a. In particular, the user has more control over which span of years should be displayed. The parameters are adjustable through an Edit window that displays the following:

**Sample numbers to print** This range field specifies the specific sample numbers to be printed.

**Start calendar year and end calendar year** These two integer fields define the horizontal axis scaling and the years that will be displayed on a graph. The default values are determined such that they will encompass all of the values of every sample in a file.

**(F)resh or (O)ven-dry conditions** As in other reports and graphs, output can be presented according to fresh F or oven-dried O conditions. The default is O.

**Show X grid (Y/N)** This asks the user if vertical grid lines should be printed across the graph at each tick on the horizontal axis. The default is Yes.

**Show Y grid (Y/N)** This asks the user if horizontal grid lines should be printed across the graph at each tick on the vertical axis. The default is No.

**Subgraphs per graph** This integer field indicates whether the data for each sample should be split into more than one graph. The default is 1 (do not split the graph).

**Subgraphs per page** This integer field allows several subgraphs to be stacked on a page. The height of each graph is adjusted accordingly. Thus, this also gives the user some measure of control on the physical size of the printed graph. The default is 3.

**Output to file or printer**

This field specifies that the output should be directed directly to the printer P or saved to a PostScript file F for later printing. The default option is P (send directly to printer). If the F option is used, a separate Edit window will request the name of the output file. This file can be printed later when a PostScript device is available, or a translator program such as GhostScript can be used to convert the file for printing to something other than a PostScript device.

## 4.7 Joining ISC Files

When ISC files are first created, the number of samples in each file is limited to the number of samples present in each particular image. It is often more practical, however, to organize samples in logical groups relating to sample site in order to facilitate comparison of samples, and production of organized reports. DendroScan allows several ISC files to be joined as long as there are no more than 1000 samples in the final file (in practice, a file with that many samples would be very large and cumbersome).

ISC files can be joined two at a time using the main menu (<**F**>ile, <**J**>oin ISC Files). A File Select window entitled Select an ISC File will open to pick the first file. Another File Select window entitled Select a Second ISC File will allow selection of the second file. The files will be joined together in a temporary file, and the user will be prompted for the name of the new file using an Edit screen entitled Save ISC File As ??. The default name that appears is always the same as that of the first file selected. The user should edit the file name accordingly.

If the name of an existing file is selected, a Verify window will declare that the file already exists, and ask if you wish the old file to be overwritten. Answering Yes will overwrite the old file with the new one just created. Answering No will allow the user to change the name of the file.

Note that if this routine is selected while editing an ISC file, the program will close it first, prompting the user to indicate whether the file should be saved before closing.

# LITERATURE CITED


Fritts, H.C. 1976. Tree rings and climate. Reprinted in 1987 in: Kairiukstis, L; Bednarz, Z.; Felikstik. E. 1987. Methods of dendrochronology (Vols. 2, 3). Int. Inst. Applied Syst. Anal. Polish Acad. Sci., Warsaw, Poland.

Gonzalez, J.S. 1989. Circumferential variation in the relative density of lodgepole pine. Can. J. For. Res. 19:276–289.

Jozsa, L.A. 1988. Increment core sampling techniques for high quality cores. FORINTEK Can. Corp., Vancouver, British Columbia. Special Publ. SP-30.

Jozsa, L.A.; Myronuk, R.M. 1986. Direct reading X-ray densitometer. FORINTEK Can. Corp., Vancouver, British Columbia. Intern. Rep.

Polge, H. 1964. Délimitation de l'aubier et du bois de coeur par analyse densitométrique de clichés radiographiques. Annales des Sciences Forestières 21:605–623.

Schweingruber, F.H.; Fritts, H.C.; Braker, O.U.; Drew, L.G.; Schar, E.; 1978. The X-ray technique as applied to dendroclimatology. Tree-Ring Bull. 38:61–91.

# APPENDIX 1
# CAPITAL COSTS AND OPERATOR TIME

**Table 1. Capital cost of setting up an X-ray densitometry system**

| Component | Estimated replacement cost | | |
|---|---|---|---|
| Sample preparation | Saw | $ 500 | |
| | Blades | 200 | |
| | Blade stiffeners | 50 | |
| | Blade separator | 125 | |
| | Jig | 100 | |
| | Miscellaneous | 200 | |
| | Subtotal | | $ 1,175 |
| Extraction and drying | Soxhlet extractor | 400 | |
| | Water extraction flask | 100 | |
| | Curved press | 120 | |
| | Miscellaneous | 200 | |
| | Subtotal | | 820 |
| X-ray and development | X-ray machine | 15,000 | |
| | 60 curved sample plates | 300 | |
| | Development tanks | 300 | |
| | Development racks (4 × 5) | 100 | |
| | Safelights | 200 | |
| | Miscellaneous darkroom equipment | 200 | |
| | Subtotal | | 16,100 |
| Paper development | Enlarger | 2,500 | |
| | Print dryer | 600 | |
| | Safelights | 200 | |
| | Miscellaneous darkroom equipment | 200 | |
| | Subtotal | | 3,500 |
| Analysis and DendroScan | Scanner | 1,200 | |
| | PostScript printer | 2,000 | |
| | Computer | 3,000 | |
| | Subtotal | | 6,200 |
| Total | | | $27,795 |

**Table 2. Operator time per X-ray sample**

| Operation | Operator time per sample[a] (min) |
|---|---|
| Sample mounting | 2.0 |
| Cutting | 2.0 |
| Extraction | 0.5 |
| Pressing | 0.5 |
| X-raying | 0.5 |
| Darkroom processing | 0.5 |
| Scanning | 1.0 |
| Preliminary data analysis | 3.0 |
| Total | 10.0 |

[a] Estimates are based on a batch of 300 short cores.

# APPENDIX 2 DendroScan OUTPUTS

## Reports

### DendroScan ASCII Format

The DendroScan ASCII report contains all of the basic information required to calculate a number of ring parameters arranged in columnar format as shown in Figure 1. This includes ring width, relative latewood width, and maximum, minimum, and average annual density. Each sample set starts with two titles, the full sample name, and a specification of whether parameters are based on oven-dry or fresh conditions. This is followed by a blank line, and a five-line series of column headers. Each data set is separated by one blank line. This report can be imported to a spreadsheet or data base through delimited ASCII filter, although the titles and column headers will not be imported. Many programs also have a file parser import option that will successfully import the entire file.

```
XLABCMP-BP-18-R1
Oven Dry Conditions

Year      ring  relative  maximum  minimum  average
         width  latewood  density  density  density
          (cm)     width   (g/cc)   (g/cc)   (g/cc)
          (OD)            (OD/OD)  (OD/OD)  (OD/OD)
----  -------  -------  -------  -------  -------
 1994   0.2415   0.5172    0.290    0.235    0.263
 1993   0.2520   0.1559    0.383    0.227    0.278
 1992   0.2226   0.4833    0.370    0.219    0.275
 1991   0.2268   0.2099    0.368    0.226    0.271
 1990   0.2247   0.5833    0.369    0.225    0.281
 1989   0.3759   0.0766    0.354    0.247    0.287
 1988   0.3801   0.2739    0.365    0.244    0.292
 1987   0.3843   0.2908    0.393    0.228    0.285
 1986   0.4053   0.1177    0.406    0.272    0.312
 1985   0.3948   0.4368    0.371    0.246    0.301
 1984   0.3108   0.1282    0.420    0.264    0.320

000000000111111111122222222223333333333444444444455
123456789012345678901234567890123456789012345678901
```

**Figure 1. Partial listing of a DendroScan ASCII Report.** This example is from an X-ray analysis, and so contains six columns. A report from a visual sample would only contain the first three columns.

## Digimic Format

The Digimic format is designed to conform to the outputs produced by the Digimic ring-width measurement system, to facilitate the conversion from this system to DendroScan. Only ring widths (oven-dry or fresh) can be reported in this format. This format includes a header line containing a number of parameters as shown in Figure 2. The header can be divided into the following fields:

| | |
|---|---|
| Characters 1–4 | Not used, zero-filled |
| Characters 5–8 | Plot code, right-justified, zero-filled |
| Characters 9–12 | Tree code, right-justified, zero-filled |
| Characters 13–16 | Sub code, right-justified, zero-filled |
| Characters 17–22 | Last year, right-justified, blank-filled |
| Characters 25–28 | Not used, zero-filled |
| Characters 31–37 | Total radius, mm |
| Characters 40–44 | Not used, zero-filled |
| Characters 49–52 | Distance to next sample, m (used for stem analysis) |

The data lines contain 11 columns. The first column (characters 1–4) contains a calendar year that marks the end of a decade. The next 10 columns

```
0000000100010001  1995  0000  0140.73  00.00    1.00
1995  00.68  00.76  00.93  01.18  00.68  01.01  01.69  01.52  01.27  01.27
1985  01.10  01.35  01.01  01.35  01.27  00.93  01.44  01.52  01.18  01.18
1975  01.01  01.10  01.18  01.77  01.86  01.44  01.44  01.77  01.10  01.35
1965  01.27  00.76  01.69  01.94  01.35  00.76  01.60  01.60  01.69  02.19
1955  01.94  01.52  01.44  02.11  02.36  02.19  02.87  01.86  02.11  02.03
1945  02.28  01.86  01.52  01.44  01.35  02.03  01.35  01.86  01.94  01.86
1935  01.60  01.94  01.77  01.77  01.69  01.94  01.94  01.94  01.69  01.60
1925  01.52  01.35  01.44  01.35  01.35  01.52  01.52  01.35  02.03  01.86
1915  01.52  01.77  01.77  01.86  01.94  01.52  01.60  01.18  01.69  01.27
1905  01.44
0000000100010002  1995  0000  0161.67  00.00    1.00
1995  00.34  00.51  00.68  00.93  00.68  01.01  01.77  01.77  01.60  01.60
1985  01.35  01.60  01.10  01.10  01.18  01.01  01.44  01.60  01.18  01.18
1975  01.10  00.84  01.01  01.77  02.19  01.69  01.69  01.86  01.35  01.44
1965  01.44  00.93  01.77  02.11  01.44  00.76  01.77  01.77  01.69  02.36
1955  02.19  01.94  01.35  02.95  02.87  02.70  05.23  03.71  02.95  02.87
1945  03.12  02.87  02.28  02.03  01.69  02.45  01.77  02.03  02.19  02.19
1935  02.11  02.53  02.36  02.19  02.11  02.36  02.03  02.11  01.94  01.52
1925  01.60  01.69  01.60  01.60  01.94  01.86  02.11  01.69  02.11  01.60
1915  01.60  01.86  01.60  01.69  01.86  01.52  01.69  00.93  01.60  01.10
1905  01.18
0000000100010002  1995  0000  0161.67  00.00    1.00
```
```
000000000111111111122222222223333333333444444444455555555556666666666777777
123456789012345678901234567890123456789012345678901234567890123456789012345
```

**Figure 2. Sample Digimic report (two samples).**

(7 characters) contain the ring width in mm for each year of the decade. The entire data set for each sample is arranged in descending calendar order (most-recent year first).

Note that the Digimic format is supposed to contain only numbers. Before creating reports in this format, make sure that the ISC file name codes (plot code, tree code, and sub code) are all numeric. Alphabetical characters will still be output to the header without error, but problems can occur when importing this file to analytical programs expecting numeric headers.

## International Tree-Ring Databank Decadal Format (RWL format)

The ITRDB decadal format was designed as a universal and compact method of transmitting and archiving raw ring parameter data (Fig. 3). It is also a format that can be imported to the ITRDB program library for further analysis in climatological studies. Data are arranged by decade, and always listed in an integer format.

All data lines start with a six digit sample name (e.g., TES001). These sample names are not directly compatible with DendroScan sample names; therefore,

```
TES001   - TESTSAMPLE-AAAA-100-AAA - PIGL
TES001   - Ring Widths (cm) X 1000
TES001  1905  1435  1266  1688  1182  1604
TES001  1910  1520  1942  1857  1773  1773  1520  1857  2026  1351  1520
TES001  1920  1520  1351  1351  1435  1351  1520  1604  1688  1942  1942
TES001  1930  1942  1688  1773  1773  1942  1604  1857  1942  1857  1351
TES001  1940  2026  1351  1435  1520  1857  2279  2026  2111  1857  2870
TES001  1950  2195  2364  2111  1435  1520  1942  2195  1688  1604  1604
TES001  1960   760  1351  1942  1688   760  1266  1351  1097  1773  1435
TES001  1970  1435  1857  1773  1182  1097  1013  1182  1182  1520  1435
TES001  1980   929  1266  1351  1013  1351  1097  1266  1266  1520  1688
TES001  1990  1013   675  1182   929   760   675  999
TES002   - TESTSAMPLE-AAAA-100-BBB - PIGL - Ring Width
TES002   - Ring Widths (cm) X 1000
TES002  1905  1182  1097  1604   929  1688
TES002  1910  1520  1857  1688  1604  1857  1604  1604  2111  1688  2111
TES002  1920  1857  1942  1604  1604  1688  1604  1520  1942  2111  2026
TES002  1930  2364  2111  2195  2364  2533  2111  2195  2195  2026  1773
TES002  1940  2448  1688  2026  2279  2870  3124  2870  2955  3715  5234
TES002  1950  2702  2870  2955  1351  1942  2195  2364  1688  1773  1773
TES002  1960   760  1435  2111  1773   929  1435  1435  1351  1857  1688
TES002  1970  1688  2195  1773  1013   844  1097  1182  1182  1604  1435
TES002  1980  1013  1182  1097  1097  1604  1351  1604  1604  1773  1773
TES002  1990  1013   675   929   675   507   338  999

00000000011111111112222222222333333333344444444445555555555666666666677777777
12345678901234567890123456789012345678901234567890123456789012345678901234567
```

**Figure 3. Sample ITRDB decadal format.**

sample names are generated at report time using the first three digits of the file name, followed by a three-digit zero-filled number corresponding to the ISC sample number.

The first two lines in a given sample can contain comments according to the ITRDB standard (Henri Grissino-Meyer, July, 1994, International Tree-Ring Data Bank, personal communication). DendroScan therefore uses these lines as data headers: the first line containing the DendroScan full sample name and the ITRDB species code; and the second line containing the data units and a multiplier for converting the integers to real numbers.

The data are arranged in ascending calendar order, in groups of 10 corresponding to even decades. The second column (characters 9–12) always contains the starting year of the decade. In the first data line, this year is equal to the earliest year in the series. In subsequent lines, it is always a number evenly divisible by 10. The data start at character position 13, and each data column is six characters wide, containing only integers. At the last data line, an extra number, 999, will appear to mark the end of data.

### Raw Profile

The raw profile report allows the entire data stream from DendroScan to be transported to another platform. These reports can be extremely large, depending on the number of samples included and the number of values in each sample. The data are divided into two columns. The first column depicts densities or gray-scale values, and the second column contains calendar year (value occurs at the ring boundary). The data are always outputted bark first, and so are in reverse chronological order. Each data set is preceded by a series of headers describing the data (Fig. 4). These include the name of the sample, the ITRDB species code, and a description line for each column. The last header entry, data spacing, specifies the distance between each sample point in cm.

### DendroScan Summary Report

The summary report is useful for data-error checking, as it contains all of the ISC header data for each sample, as well as a number of calculated values that are otherwise unavailable. The data are arranged in two tables (Fig. 5). It is recomended that this report be produced and archived at the end of all analyses. The reports are especially useful when recalibrating X-ray wedges because they list the average sample density. They are also useful when estimating the ages of trees from cores.

```
SampleName  = TESTPINE-1-1-1
Species     = PCGL
Column_1    = Density g/cc (Oven_Dry)
Column_2    = Boundary_Year
DataSpacing = 0.00844 cm (Oven_Dry)

 0.3989  1995
 0.3681
 0.3487
 0.3365
 0.3289
 0.3289
 0.3368
 0.3653  1994
 0.4011
 0.4050
 0.3728
 0.3445
 0.3259
 0.3092
 0.3060
 0.3078
 0.3237  1993
 0.3533
 0.3881
 0.3780
 0.3457
 0.3223
 0.3071
 0.2952
 0.2914
 0.2974
 0.3116
 0.3455  1992
 0.3835
```

```
00000000011111111112222222222333333333344444444445555555555
12345678901234567890123456789012345678901234567890123456789
```

**Figure 4. Raw profile report.**

```
                                ITRDB   ITRDB                    Samp  Raw Dry Pixel
    Sample ID----------- Sample  SPP   1st  Last Total Hght  Radius  Resol
Num Stand      Plt Tr Sub Code    Code  year year Rings m     cm      mm
--- -------------------- ------- ----- ---- ---- ----- ----- ------- -----
  1 TESTPINE   1    1  1  TES001  PCGL  1905 1995    91  1.30  14.073 0.084
  2 TESTPINE   1    1  2  TES002  PCGL  1905 1995    91  1.30  16.167 0.084

    Centre Adj Dry Est   Fresh   Xray   Dry   Dens   Basic
    Missed Radius  Age   Radius  Thick  Dens  Fact   Dens
Num cm     cm      yrs   cm      cm     g/cc         g/cc
--- ------ ------- ----- ------- ------ ----- ------ -----
  1  0.000  14.072    91  16.000 0.1420 0.402 0.9010 0.362
  2  0.000  16.167    91  16.000 0.1420 0.378 0.9010 0.341
```

```
00000000011111111112222222222333333333344444444445555555555666666666677777
12345678901234567890123456789012345678901234567890123456789012345678901234
```

**Figure 5. Sample DendroScan summary report.**

## GRAPHICS OUTPUTS

Graphs can be produced directly from DendroScan on a PostScript device. Outputs can be many pages long (and so do not conform to Encapsulated PostScript convention). Each page contains a header line specifying the type of graph, the date it was produced, and a page number (*see* Figs.6 to 9). Four general graph types are produced.

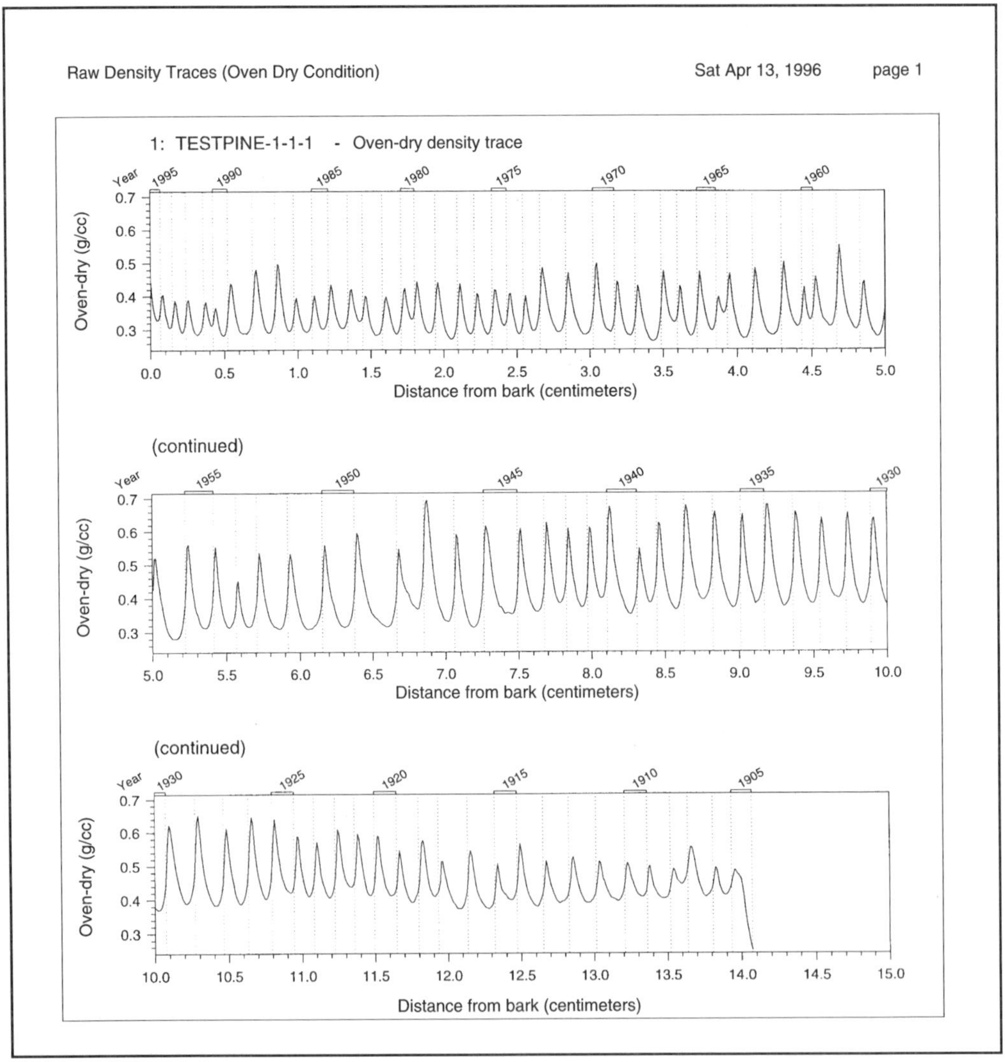


**Figure 6. Example of a density profile graph.**

## Raw Density Profile

The density profile graph is the most complex graph produced by DendroScan and is similar to the graph displayed in the ISC Edit screen. Figure 6 illustrates a profile of a single sample split into 5-cm segments. The X axis represents distance from the bark in cm, and the Y axis represents density or gray scale (depending on the sample type). The vertical dotted grid lines define annual ring boundaries. Years are marked at the top of the graph (in intervals of five). If graphs are produced for several samples at once, each sample will appear on a different page. These graphs are useful for displaying the completed analysis for archiving or publication.

## Annual Density Profile

Annual density profiles are similar in appearance to the raw density, but they differ in several ways. The density profile for each year has been normalized so that it stretches accross the entire year, and the X axis is divided into calendar years (Fig. 7). Note that because annual density profiles are plotted against year, they are backwards to the raw density profiles.

Annual density profiles are particularly useful when attempting to compare samples based on intra-year details such as frost rings, false rings, or other anomalies. It is often useful to limit the span of calendar years printed so that details are easily discerned in individual years.

## Sample Summary Graphs

Summary graphs are a predefined layout of three graphs per sample, arranged on a single page (Fig. 8). The top graph displays ring widths; the middle graph, density (maximum, minimum, and average); and the bottom graph, relative latewood width.

## Single Parameter Graphs

Single parameter graphs (ring width, density, relative latewood width) are useful in comparing a single parameter for a number of samples (Fig. 9). All graphs in a single output are plotted against the same calendar period, allowing rapid comparison. They can be stacked as high as five graphs per page.

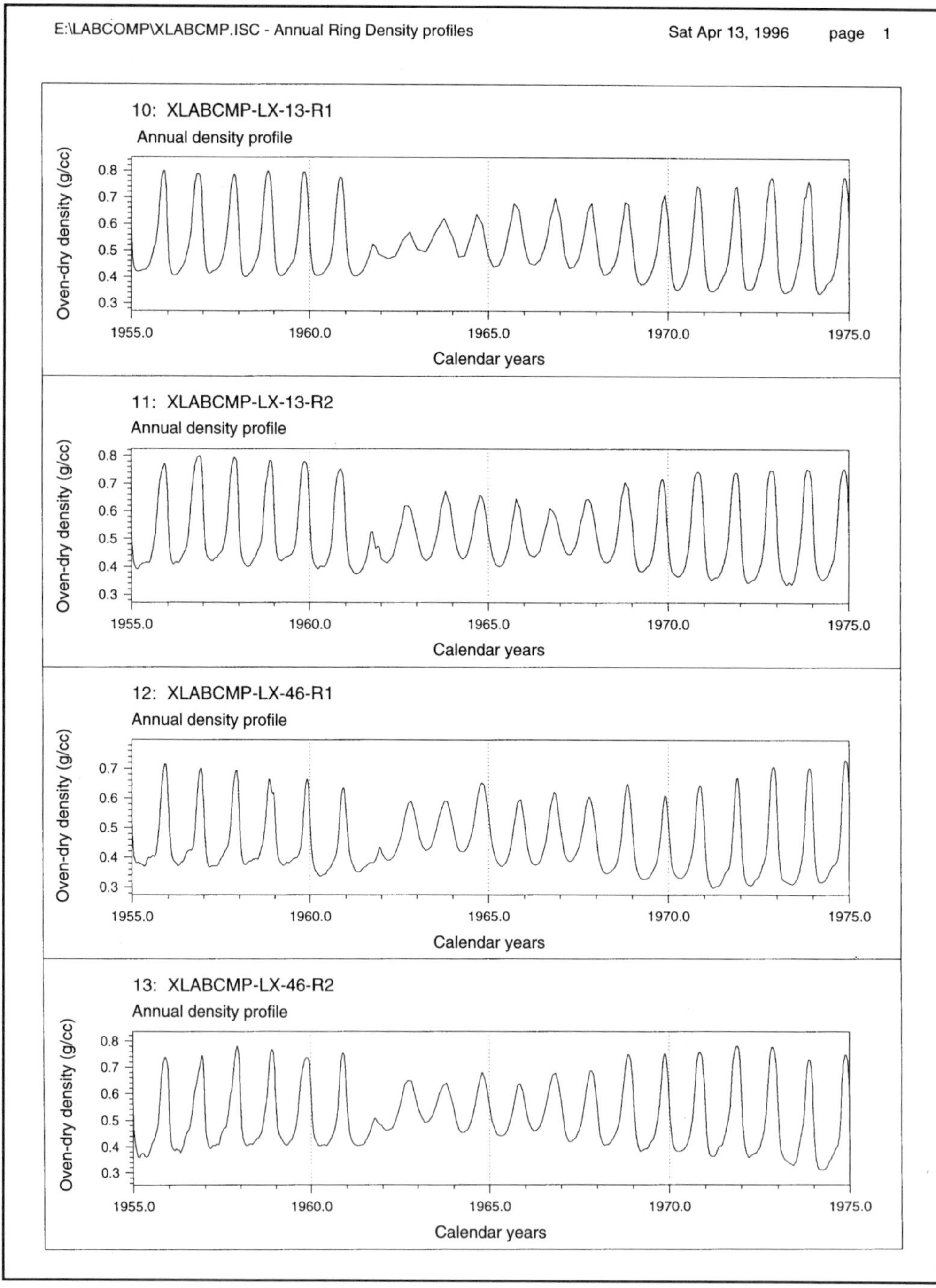


**Figure 7. Sample page of annual density profiles (4 samples).**

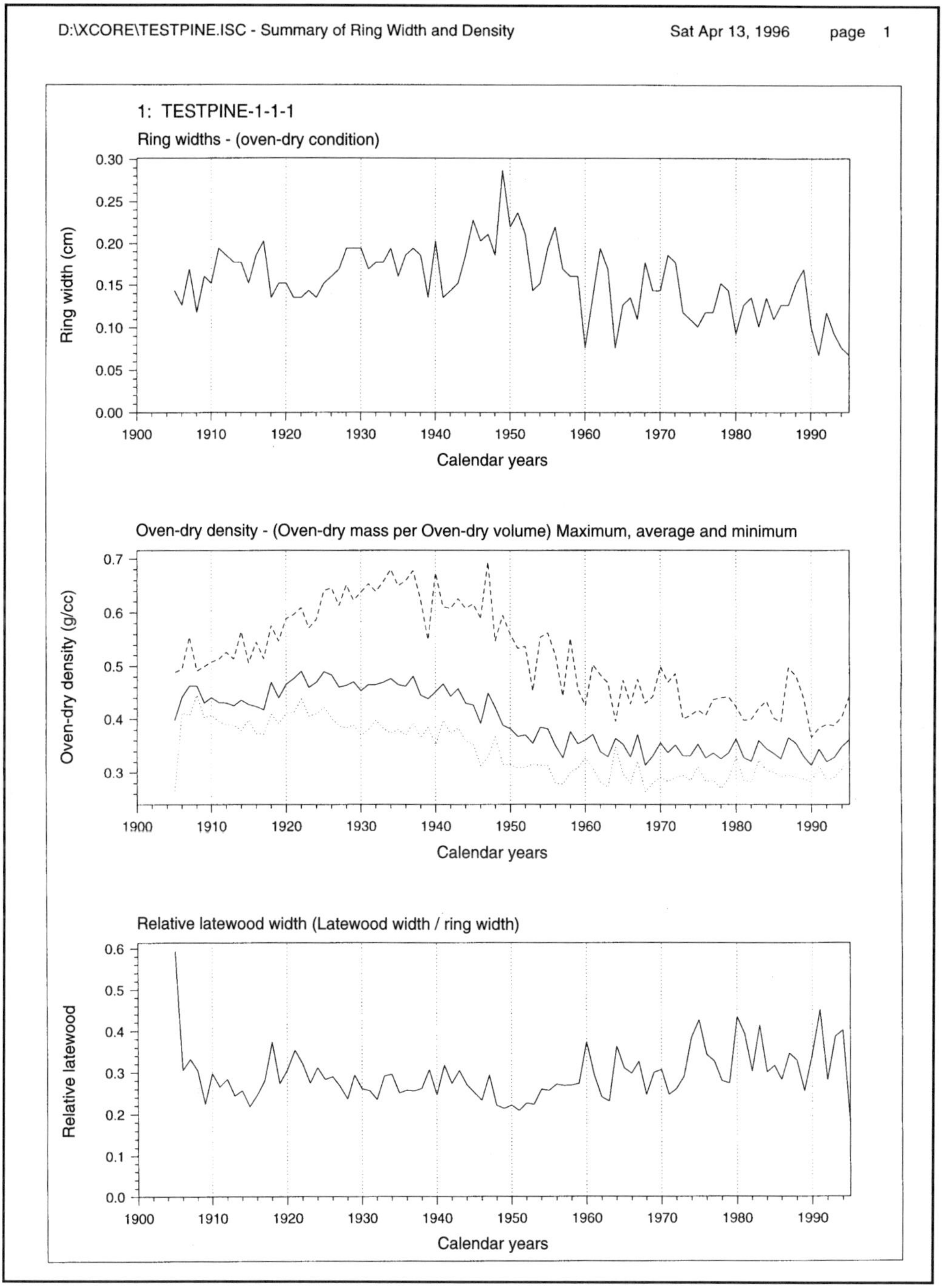


**Figure 8. Sample summary graph page.**

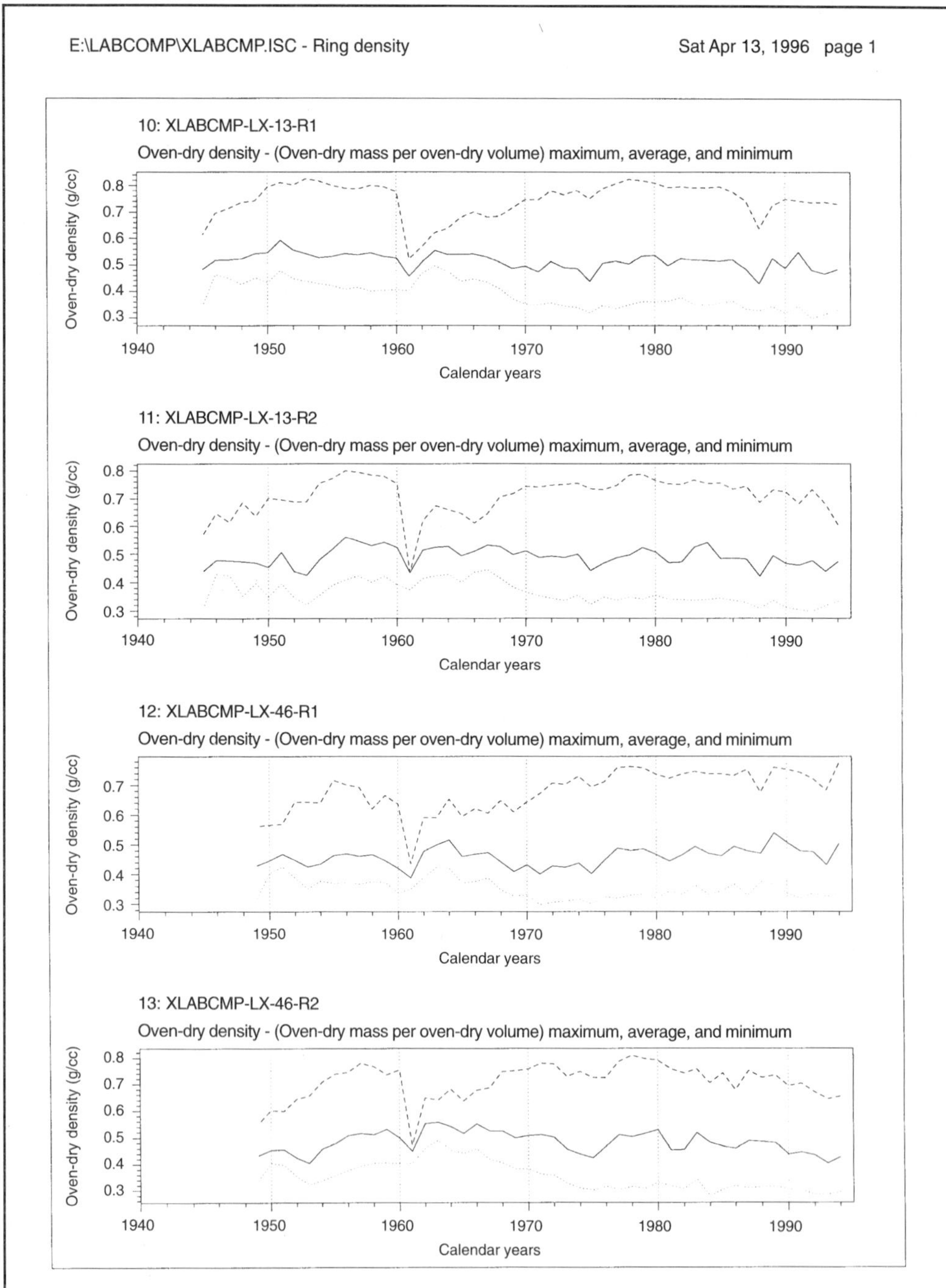


**Figure 9. Sample page of single parameter graphs.**

# APPENDIX 3 DENDSCAN.INI

The file DENDSCAN.INI contains information and settings vital to the correct operation of DendroScan. The DENDSCAN.INI file can be modified using an ordinary text editor. The following rules must be followed:

- Only TEXT characters and spaces are allowed. Control characters and tabs should not be used.
- Comments can be included in the file, but they must be enclosed in curly brackets { }.
- Upper- and lower-case characters can be used, as they are all converted to upper-case when read by DendroScan.
- Section header labels must be enclosed by square brackets [ ].
- Spaces are allowed anywhere except where they would break up a variable or header label.
- Spelling must be precise when specifying header or variable labels.
- Only one variable or header label can be included on each line, although some variables take two values.
- If a variable label takes two values (e.g., wedge data), the values must be separated by a single comma (although blanks are allowed with the comma).
- All values must be separated from the variable label with an equal (=) sign.

The file is divided into two sections, which must each be identified by a section header label. The first section is labeled the [Basic info section] and contains the following variables.

**Minimum_Path_Pixels = 20**

When TIF files are created by a scanner, occasional digital errors occur and a random black dot might appear. This defines the minimum number of consecutive black dots that will identify a sample on the picture. Twenty is a good value, as it allows great flexibility when marking samples with relatively short segments, yet it fulfills the job of missing stray pixels. The maximum value allowed is 25 and the minimum is 5.

**Pixel_Bandwidth = 5**

When DendroScan averages densities or gray values, it does so along an imaginary band that extends perpendicular to the marked black line on either side of it. This variable defines how long this band will be (in pixels). The shorter the band, the more accurate the density measurements. A longer band could be useful to produce more repeatable measurements in very heterogeneous wood such as diffuse porous angiosperms. The maximum value allowed in this location is 20, and the minimum is 4. As a rule, if mostly gymnosperms are measured, this value should be left as 5.

**Wedge_Number = 1**

Different calibration wedges can be used for different purposes, and DendroScan allows the user to maintain data for up to five calibrated wedges; however, there is usually one wedge commonly used for most wood samples. Wedge_number simply allows the user to specify a default wedge number. This can always be overridden during operation of DendroScan prior to analyzing a TIF file.

**Sample_Thickness = 0.150**

Depending on the equipment used for cutting samples, a different sample thickness can occur. This value allows the user to specify the default thickness most commonly used with the equipment used. This specification is in cm. The user will always be given the option to override this value when analyzing a TIF file.

The second section is called the [Wedge section]. When wedges have been calibrated, their calibration data may be entered here.

The wedge section contains information about five wedges. Each wedge starts with the header "Wedge_X" (where "X" equals a number from 1 to 5) and ends with the footer "end_wedge". If a wedge has been calibrated, it will contain several values, which are outlined here.

**Serial_Number = 12044** {reference purposes only}

This is the serial number of the wedge. This number is for reference purposes only. This number should be the same as the number of the wedge included with this package.

**Calibration_Constant = 0.6621 {calculated for 0% moisture}**

This is a calibration multiplier used to convert wedge thickness into equivalent wood density (*see* Appendix 4).

**Wood_Moisture_Percent = 6.0 {based on an RH of 30%}**

This number is a percent value. DendroScan uses this value to adjust the calibration constant so that reported densities are oven-dry. This number is initially set during installation, but can be changed later. Appendix 4 provides a graph for selecting the correct wood moisture percent based on the relative humidity of the darkroom.

**Reference_Length = 1.5310** {length of reference edge}

This is the length of the reference edge on the wedge. This length is critical, as it is used by DendroScan to scale the size of an image.

**Reference_Offset = 0.091** {distance reference edge to main edge (cm)}

This refers to the distance that the reference edge juts out from the main body of the wedge.

**Distance_to_Midline = 0.294 {width distance from the reference edge to the midline}**

The midline is an imaginary line along which DendroScan reads gray values from the wedge. This line is parallel to the reference edge. The Distance_to_Midline is the distance between the reference edge and the midline.

**Wedge_Length = 2.5353 {length of the wedge if it tapered to zero (cm)}**

This is not the actual length of the wedge, but rather the ideal length if the slope continued to a razor-sharp tip. In practice, the tips of the wedge are slightly shorter due to material limitations. (It is normal for the wedge to be slightly chipped at the tip.)

**Cutoff_Length = 2.4966** {physical length of the wedge (cm)}

It is not possible for the wedge to taper to a zero thickness at the tip. Therefore the wedge is usually physically shorter than the distance defined in Wedge_Length above. Typically the install routine sets this value as 0.1 cm shorter than Wedge_Length. If the tip of the wedge becomes damaged, this value should be adjusted accordingly.

**Wedge_Height = 0.5113** {thickness of the wedge at the thick end (cm)}

This is the thickness of the wedge at the thickest part.

**Minimum_Thickness = 0.01** {minimum thickness used for calibration (cm)}

This specifies the minimum thickness (at the tip) where the wedge is considered reliable. DendroScan will not measure gray values at the tip beyond this point. Typically, the value is 0.010 cm, which for a 1.5-cm-thick wood sample, works out approximately to a density of 0.04 g/cc. Beyond this point, the tip of the wedge is sometimes chipped or otherwise damaged.

# APPENDIX 4 WEDGE CALIBRATION

## Mathematics

X-rays interact with matter in complex ways, but the absorption of low-energy X-rays of a specific wavelength is proportional to the thickness and density of a given substance through which the X-ray passes, as per the following formula (Agfa-Gevaert N.V. 1986. Industrial radiography. Agfa-Gevaert N.V., Mortsel, Belgium.).

$$\boldsymbol{A = L \times T \times D}$$

$A$ = X-ray absorbance, which is defined as the change in X-ray intensity open after passing through a sample divided by the initial intensity (a unitless value from 0 to 1)

$L$ = The linear absorption coefficient of the material, which is a function of the composition and specific weight of the material ($cm^{-1}$)

$T$ = Thickness of the material (cm)

$D$ = Density of the material (g/cc)

In the case of wood, the thickness can be controlled and the density is unknown. In the case of the calibration wedge, the density is a constant and the thickness is known. For any specific gray value in an image, the absorbance is fixed. Thus, for any specific gray value, the density and thickness for a calibration wedge relate to the density and thickness of a wood sample as follows:

$$\boldsymbol{L_w \times T_w \times D_w = L_c \times T_c \times D_c}$$

Subscript *w* refers to a wood sample

Subscript *c* refers to the calibration wedge

The absorption coefficients and the wedge density are all constant, so this equation simplifies to the following:

$$\boldsymbol{D_w = (T_c \times K) / T_w}$$

$K$ = Calibration constant—a constant relating the density of the wedge to the density and thickness of a wood sample

The thickness of a wedge ($T_C$) is known in detail. The thickness ($T_W$) of the sample can be measured. The density of the wood sample can therefore be calculated if the calibration constant ($K$) is known.

## Determination of a Calibration Constant

The calibration wedge included with this manual is made of epoxy resin. It has been precalibrated at a relative humidity of 30% and the calibration constant was determined experimentally to be 0.66 against a number of extracted samples.

Verifying this constant can be conducted using the following procedure:

- Prepare several samples as described in Part 2 of this manual.
- Cut the samples to a length of about 3 cm and a width of about 0.5 cm. Make the cuts as precise as possible so that the width is constant along the length of the samples and the ends are squared-off. Then measure the width, length, and thickness of the samples with precision calipers to obtain a sample oven-dry volume in $cm^3$. (This is actually the air-dry volume, but for all practical purposes, this volume will be the same as oven-dry.)
- Store the samples in a humidity-controlled darkroom for a period of 24 hours, and weigh them on an analytical balance. (Keep the humidity constant throughout this operation.)
- Correct the sample mass to oven-dry using the information in Figure 1. (For example: a sample weighs 0.7500 grams at a relative humidity of 30%. Figure 1 shows that the sample should contain 6.0% moisture. The oven-dry mass for this sample is therefore 0.7500 divided by 1.06, or 0.7076.)
- Calculate the oven-dry density for the sample by dividing the oven-dry mass by the measured volume.
- X-ray, scan, and run DendroScan for these samples (include the calibration wedge in the image). Output a summary report from DendroScan. This report will include the calculated average dry densities of the samples according to the current calibration constant.
- The calibration constant correction can be adjusted by dividing the gravimetric oven-dry density by the reported oven-dry density, and averaging

this number for all samples. A new calibration constant is calculated by multiplying the old calibration constant by the average correction.

- To change the calibration constant for DendroScan, simply change the value for calibration_constant in the appropriate wedge section in the DENDSCAN.INI file.

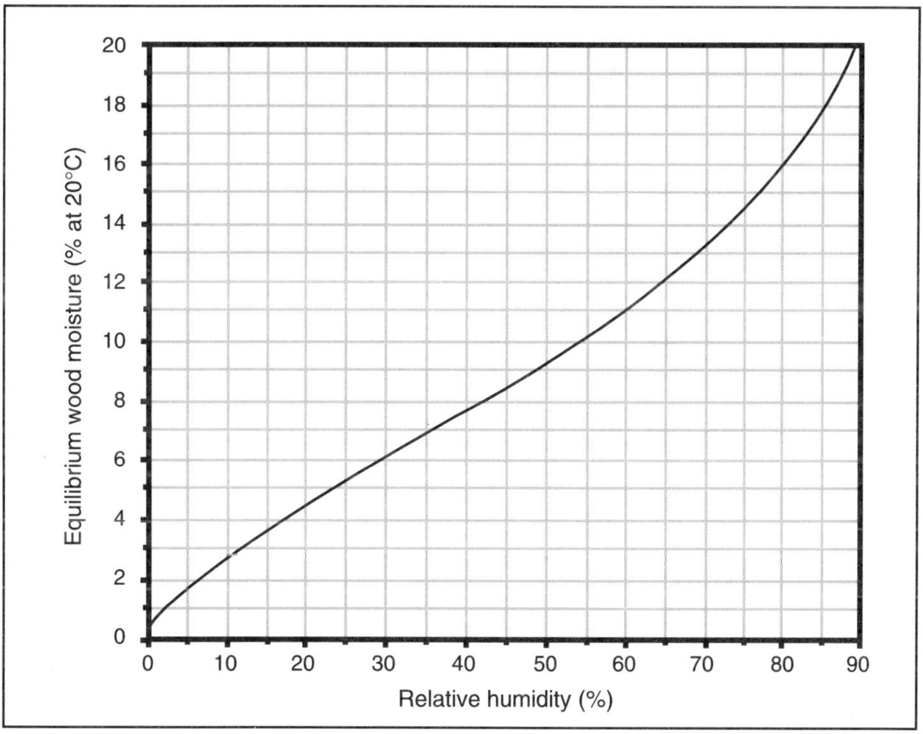


**Figure 1. Predicting moisture content of wood samples** This curve was derived using the method described by Zuritz et al. (1979)[a] and paramaterized by Avramidis (1989)[b]. The curve appears to be equally valid for both raw and extracted wood samples.

[a] Zuritz, C.; Singh, R.P.; Moini, S.M.; Henderson,S.M. 1979. Desorption isotherms of rough rice from 10°C to 40°C. Trans. ASAE 22:433–440.

[b] Avramidis, S. 1989. Evaluation of "three-variable" models for the prediction of equilibrium moisture content in wood. Wood Sci. Technol. 23:251–258.

## Pseudowedges

Some wood species differ somewhat in the degree to which they deflect X-rays (Schweingruber et al. 1978).[1] In this case, it is useful to create pseudowedges. These are only basically a description of the same wedge, but where the calibration constant is slightly different. For X-ray purposes, the same physical wedge can be used. When DendroScan is being run, however, a different wedge number is used. To create the data for such a wedge, simply copy the wedge information from wedge_1 in the DENDSCAN.INI file to another wedge number using an ordinary text editor and then edit the calibration factor for the new wedge.

## Recalibrating for Different Relative Humidity

When DendroScan is first installed, it installs a wood moisture percent value in the DENDSCAN.INI file. This value assumes that the relative humidity of the darkroom is constant (to ±5% relative humidity) and that the wood samples have been allowed to equilibrate with the darkroom environment for several hours. If for some reason, the relative humidity controls in the darkroom need to be changed, the percent wood moisture can be easily adjusted using Figure 1. For example, if the relative humidity is 30% and it will be changed to 40% simply edit the DENDSCAN.INI file and change the Wood_Moisture_Percent value. The old value is 6.0, and should be changed to 7.6.

## Broken Wedge Tips

Even with careful handling, the tip of the wedge can sometimes become damaged. Before replacing the wedge, measure the thickness of the tip of the wedge. If it is thinner than 0.015 cm, then it will still function satisfactorily. Even if it is slightly thicker (because the break was larger), this can be adjusted by using a text editor to change the Cutoff_Length value in the wedge section of the DENDSCAN.INI file to the thickness at the edge of the break. This should not significantly affect density estimates as long as the minimum thickness is less than 0.035 cm (for sample thicknesses of about 1.5 mm).

1 Schweingruber, F.H.; Fritts, H.C.; Baker, O.U.; Drew, L.G.; Schar, E. 1978. The X-ray technique as applied to dendrochronology. Tree-Ring Bull. 28:61–91.

# APPENDIX 5 TROUBLESHOOTING

DendroScan can display a number of error messages, which usually are displayed when a potentially fatal error is detected. This appendix lists all of the error messages that can be encountered, as well as a number of problems that the user could experience. This appendix is divided into sections corresponding to the DendroScan activity being conducted when the problem or error occurred. If you discover a problem, look it up in the appropriate section. If an message/warning window accompanied the problem, note the number in parentheses in the message box header and look it up in this section. The error messages are listed in numerical order.

## Opening a TIF File

The TIF format is complex and can contain many types of images. Unfortunately, DendroScan can only deal with 8-bit, uncompressed, gray-scale formats. To avoid system crashes, the TIF file header is checked before the file is processed. If a problem is discovered, a descriptive message will appear along with the standard header line numbers (Kay and Levine 1992)[1] where the error was encountered. If the error is too severe, DendroScan will refuse to open the file.

**MESSAGE (1) Invalid TIF format**

TIF files contain a specific label at the start of the file. If this label is missing, DendroScan assumes that the file is not a correct format and will refuse to load it. This error occurs when some other file type has been selected by mistake.

SOLUTION: Check the file name to see that it was entered correctly. Alternatively, if it happens often, the problem probably lies in the scanning software.

**MESSAGE (3) Image not 8-bit [258]**

DendroScan will refuse to load a file whose pixels are not encoded as 8-bit.

SOLUTION: Check the file attribute preferences in the setup of the scanning software to ensure that the file is both scanned and saved in 8-bit format.

[1] Kay, D.C.; Levine, J.R. 1992. Graphic file formats. First edition. Windcrest/McGraw-Hill (Tab Books), Blue Ridge Summit, Pennsylvania.

**MESSAGE (5)  Compressed image [259]**

Compressed images are not allowed.

SOLUTION: Check the scanning software setup routine to ensure that the file is saved in uncompressed format.

**MESSAGE (7)  Color image [262, 277, 284]**

Only gray-scale images are allowed.

SOLUTION: Check the scanning software setup routine to ensure that files are scanned in gray scale only.

**MESSAGE (9)  Missing image locaters [256, 257, 273]**

DendroScan could not locate the width or height (pixels) of the image or the file location of the image.

SOLUTION: Check that the correct file was opened. Rescan the image if necessary.

**MESSAGE (11)  Unstated resolution [282, 283, 286]**
******* Assumed at 600 dpi *******

Most scanning software allows a choice of whether the image dimensions are recorded as inches, centimetres or pixels. This error occurs when pixels have been selected.

SOLUTION: If the sample is an X-ray, DendroScan will recalculate the image resolution from the calibration wedge. The problem can be solved for all types of images by ensuring that the image is saved with in. or cm as dimensions.

Note: The complete structure of a TIF file can be examined using a utility called TIFCHECK (included with the DendroScan program files). The file can be run from DOS with the command:

**TIFCHECK** (file name)

This produces a full-screen output similar to the one displayed in Figure 1. If TIF file structure errors persist, the output from this file can be used to discuss the problem with the authors.

```
------------------------- HEADER TAGS - D:\XCORE\test.tif
(254) New Subfile Type = 0
(256) Image Width = 1086 columns
(257) Image Length = 302 rows
(258) Bits Per Pixel = 8
(259) Compression = 1 (no compression)
(262) Photometric Interpretation = 1 (GreyScale - 0 = BLACK)
(271) Scanner/Software = MICROTEK
(273) Picture Start Location = 280
(277) Number of Colour Planes = 1
(278) Rows Per Strip = 302 (1 strips)
(279) Bytes Per Image Strip = 327972
(282) Picture Width resolution =   200.00
(283) Picture Height resolution =   200.00
(284) Planar Configuration =1 ( GreyScale )
(286)
(287)
(296) Resolution Unit = inches
------------------------- SUMMARY - D:\XCORE\test.tif
COLOUR:          8 bit GreyScale (BLACK = 0)
RESOLUTION:      200.0 pixels per Inch
DIMENSIONS:      1086 X 302 pixels  (5.43 X 1.51 Inches)
SOFTWARE:        MICROTEK
```

**Figure 1. Sample output from the TIFCHECK utility.**

## Translating a TIF file

### MESSAGE (12) Short line segment found and ignored

DendroScan found a path line that was shorter than that specified in the DENDSCAN.INI file (Minimum_Path_Pixels). This error is not fatal, but it should be taken seriously. The result is that the program will not follow the entire path indicated by the user, and part of the sample will be missing. Note: to avoid system crashes, DendroScan will never attempt to read a line that is completely vertical or that contains only one pixel.

SOLUTION: Stray black pixels can be alleviated by raising the brightness controls slightly when scanning an image. This ensures that stray black pixels do not occur. Ensure that path lines are not vertical in the image.

### PROBLEM Several samples apparently linked before the linking procedure was started (a blue box surrounds several samples)

If the linking procedure has not yet started, then linked samples indicate poorly drawn path lines and no clear horizontal gap between the samples.

Even experienced users can accidently generate this problem if the samples are tightly packed in the image.

SOLUTION: Part 3 of the manual describes the correct procedures for drawing lines. The simplest way to avoid overlapping samples is to ensure that the samples are arranged as close as possible to horizontal, and that a white space exists between each piece.

**PROBLEM The red lines overlay the blue lines for a while, and then appear to cross between the blue lines**

This problem also relates to poor path layout and occurs when there is no clear vertical gap between different line segments in a sample. This can be a serious error. The red line is the feedback that tells how well the paths were translated by DendroScan. When a red line jumps between the blue lines, the image is also read across this gap, resulting in extraneous data in the sample.

SOLUTION: Edit the image lines to fix the problem, or rescan and redraw the lines.

**PROBLEM A single sample appears to have been split into two**

If there is a horizontal gap between line segments within one sample, they will be translated as separate samples.

SOLUTION: If the pieces are consecutive, they can be linked during the linking procedure. Otherwise, if one of the samples appears to be part of the middle of the other sample, the TIF file will have to be re-edited, and an elevator line drawn (*see* Part 3 of this manual), where it will ensure that the segments are all seen as part of the same sample.

**MESSAGE (13) Wedge number # is not calibrated**
**The file will not be analyzed as an X-ray**

This message will occur immediately after the wedge number has been selected. Either the wrong wedge number was selected, or the wedge selected has not been calibrated.

SOLUTION: If the wrong wedge number was inputted, stop the scan by pressing any key, and reload the file. Alternatively, check the DENDSCAN.INI file to see if the wedge selected actually contains calibration information.

**MESSAGE (15) Cannot link samples # and #**

Samples can only be linked if they run consecutively in the file. This message occurs if the sample numbers were not consecutive (e.g., 5 and 9) or if the same sample was selected for both prefix and suffix (e.g., 1 and 1).

SOLUTION: If you need to link the nonconsecutive samples, reorganize the image to make them consecutive, then rescan and reline.

**MESSAGE (17) Cannot complete link**
**Cannot open ISC file**

This usually indicates a system problem and is more likely to occur when running DendroScan from Windows. It indicates that either the temporary files have become corrupted due to a disk error, or another program has attempted to open the same files.

SOLUTION: Close all other programs if running in Windows or reboot the system. The problem should not repeat on a second attempt.

**MESSAGE (19) Problems creating ISC**

This error message can occur just after the TIF file translation has been completed. It usually indicates a system problem similar to Message 17.

SOLUTION: Close all other programs if running in Windows or reboot the system. The problem should not repeat on a second attempt.

## Editing an ISC File

**PROBLEM DendroScan picked ring boundaries at too many locations**

The algorithm used by DendroScan to select ring boundaries uses both the shape and relative height of the density signal. It also makes an assumption that the density of noise is less than 30% of the density of the sample (signal-to-noise ratio of 3 or higher). Greater noise will occur in samples that have not been properly extracted, or in diffuse porous woods, which have large random deviations in the density signal even within the ring.

SOLUTION: Ensure that samples are completely extracted. The noise level in diffuse porous woods can be reduced by increasing the scan width (adjusting the pix_averaging parameter in the DENDSCAN.INI file from 4 to 20) and then retranslating the TIF file. If ring boundaries are still difficult to discern from noise, viewing the X-ray film on a microfiche viewer during analysis can help.

**PROBLEM One of the rings in the sample is an unusual shape, with the appearance of a shelf with a lot of random noise**

This problem is usually related to inexperienced drawing of path lines in the original TIF file. It is caused when there is not a vertical gap between different line segments in a sample. When DendroScan translates these segments, it analyzes an imaginary line between them. The resulting signal is a very noisy, flattened ring-like structure in the density signal.

SOLUTION: Check and adjust the line drawing in the TIF file, and repeat the TIF translation to create a new ISC file.

**PROBLEM Many rings appear flattened**

Another cause of a flattened density signal is when the rings in the original sample are poorly aligned. This usually results in many flattened rings with very little noise. Furthermore, many rings are usually affected. An examination or the X-ray image can verify this problem, as it results in an apparent blurring.

SOLUTION: Ensure that samples are collected and prepared with care.

**PROBLEM DendroScan will not allow a ring marker to be set**

This can occur if the sample pointer is located in a density valley. If this is the case, a lower-case v will appear on the bottom left of the screen on the status line.

SOLUTION: Moving the pointer to the next pixel (using control and the right or left arrows), allows the boundary to be placed there.

**MESSAGE (31) Invalid ISC structure**

This message indicates that the user has tried to open a file that does not contain the correct label for the ISC format. This could be due to accidentally selection of the wrong file, or can be related to hard-disk problems.

SOLUTION: Check that the correct file was selected, or rebuilt the ISC file from the original image.

**MESSAGE (33) Problems creating temp file**

This error message can occur when opening an ISC file. DendroScan always makes a copy of the file to a temporary file in its own directory. This error can occur if the disk where DendroScan resides is full.

SOLUTION: Delete extraneous files to free up disk space.

**MESSAGE (35) 3 blank years here already
I cannot add another!**

DendroScan only allows a maximum of three blank (missing) rings to be marked at any given ring boundary.

SOLUTION: If more than three blank rings need to be placed at one location, this can be done by adding a new ring boundary very close to the original, resulting in a very thin ring. This is not quite the same as adding a blank year, but has nearly the same effect. Furthermore, more blank years can also be added at this new thin year.

**MESSAGE (37) Blank years can only be added
where a current year marker exists**

This message appears if the user attempts to insert a blank year in the middle of a year. This is not possible, because it would cause the existing year to be split into two.

SOLUTION: Move to a ring boundary to add blank years.

**MESSAGE (39) No blank years here
I cannot remove one**

SOLUTION: Move the pointer bar to the boundary line before removing the blank years, or use the space bar to remove the ring boundary if that was the intention.

**MESSAGE (41) You cannot delete the only sample in this file,
delete the file instead**

It is possible to delete samples from a file (as described in Section 4.4.g); however, an ISC file by definition must contain at least one sample.The program will therefore not allow a user to delete the last sample from a file.

SOLUTION: Delete the file.

**MESSAGE (43) Cannot save ISC file**
**Temp file is not open**

This error message is unlikely to be encountered. It either indicates that another program attempted to access DendroScan's temporary files, or there was a disk failure while operating DendroScan.

SOLUTION: Close all other programs if running in WINDOWS or reboot the system. The problem should not reoccur on a second attempt.

**MESSAGE (45) Cannot save ISC file**
**Cannot open ISC file**

This message can occur if a user is editing an ISC file on a floppy disk. If the disk is removed from the drive while DendroScan is operating, attempting to save the file will yield this error.

SOLUTION: Reinsert the floppy disk and try again.

## Edit Fields

Wherever edit fields are used in DendroScan (in Edit windows and the Spreadsheet window), checks are made before a user input is accepted. The following problems and messages can occur while inputting information on the screen.

**PROBLEM Beeping noise**

Placement of characters in fields is restricted. Attempting to enter an invalid character results only in a beeping noise.

SOLUTION: Section 4.2.g details the legal characters allowed in different types of fields.

**MESSAGE (51) Bad integer format**

This message occurs when an input can not be translated as a valid integer (for example, a 1 followed by a minus sign).

SOLUTION: Reinput the correct integer.

**MESSAGE (53) Bad real format**

This message occurs when an input can not be translated as a valid real number (for example, two decimal points in a row).

SOLUTION: Reinput the correct real number.

**MESSAGE (55) Out of range (# to #)**

DendroScan also limits the range of values that can be inputted into real, integer, and range fields. The error message lists the allowed range. For example, the allowed value for sample thickness ranges from 0.01 cm to 1.0 cm. Entering a number higher or lower than this value into this field will trigger an out of range error. When this type of error occurs, the program will not allow the user to leave this field until it receives a valid entry.

SOLUTION: Reinput the value, ensuring that it is within the range indicated in the error message.

## Producing Outputs

**PROBLEM Graphs do not print**

This is due to several reasons. DendroScan is only capable of sending output to the first parallel port (LPT1). In some networks, the printer is connected to some other logical port. In this circumstance, graphs cannot be printed.

SOLUTION: Adjust the system configuration so that the printer is connected to LPT1.

**PROBLEM Instead of graphs, the printer produces many pages of incomprehensible text**

This text output is probably the PostScript code that defines the graph. The most-probable reason for this problem is that the printer is not a PostScript device, or the printer is not in PostScript mode.

SOLUTION: Check the printer manual to ensure that your printer is a PostScript device, and that it is set correctly. If you have no access to a PostScript printer, you can aquire a commercial translator (e.g., GhostScript) to print the files on a nonPostScript printer. Note that Dendroscan outputs are not Encapsulated PostScript and so cannot be imported as EPS files. As a last alternative, DendroScan can produce an ASCII report, which can be imported to most popular graphing software.

**PROBLEM Cannot control the 'Y' axis (density) range in graphs**

DendroScan automatically scales the Y axis on all graphs it produces. This facilitates the rapid production of many graphs for files that contain many samples, and simplifies the programming. If different Y scales are required for a publishable quality graph, a DendroScan ASCII report can be produced and imported to a spreadsheet or other graphing package.

## DendroScan Shutdown

**MESSAGE (61) Out of Memory** (Followed by a shutdown)

DendroScan operates in the DOS memory area, which is limited in total memory to 640 kB. It requires at least 450 kB of this memory to operate. This message can occur anywhere in the program if a shortage of memory is encountered. Immediately after this message, DendroScan will shutdown.

SOLUTION: If a shutdown with this message occurs, check the DOS memory using the command: mem/C|more. This will summarize the TSRs (terminate and stay resident programs) present. Unloading any nonessential TSRs will usually free up a considerable amount of memory.

**Recovering from a shutdown**

Dendroscan maintains a temporary file when the user is editing an ISC file. This file is dynamically maintained during an edit operation (every time the user switches between samples). Thus, if a system shutdown occurs during an ISC edit operation, most of the editing session can be recovered. First, fix the problem that caused the shutdown (see above). Then locate a file called TEMP.ISC in the DendroScan directory. This file is a copy of the last edit session. To recover the session, copy this file to a new directory and rename it. You will find that most of the changes made to the samples in the file have been preserved.

# APPENDIX 6
# EDITING SPECIES LISTS AND HELP FILES

Species lists and help files are ASCII files with a fixed line-length. They can be easily edited with any text editor (DOS Edit, WINDOWS NotePad, etc.). The fixed line-length means that they can also be handled as random access files that can be searched backward and forward. This makes them particularly useful for help files and lookup files that need to be rapidly searched for information. Figures 1 and 2 illustrate this structure.

The first line of a list or help file contains a command that is utilized by DendroScan: ListWid = # (where # refers to a maximum line width). The maximum line width is usually set to 73, which is the maximum width in characters that can be comfortably displayed on a VGA graphics screen. This number should never be adjusted, and this command must always appear at the start of the file. When editing the file, data should not extend beyond column listed in this statement. The vertical bar at the end of each line located at the next column to that specified by ListWid, and is a reminder of the maximum line length. When a list file is being edited, these bars do not need to be inserted manually; this function is automatically performed by a DendroScan utility program called FIXLST.

```
ListWid = 73                                                              |
ABAM  Abies amabilis                      Pacific silver fir              |
ABBA  Abies balsamea                      balsam fir                      |
ABCO  Abies concolor                      white fir                       |
ABFR  Abies fraseri                       Fraser fir                      |
ABGR  Abies grandis                       grand fir, giant fir            |
ABLA  Abies lasiocarpa                    subalpine fir                   |
ABMA  Abies magnifica                     California red fir              |
ABPN  Abies pinsapo                       Spanish fir                     |
ABSI  Abies sibirica                      Siberian fir                    |
ABSP  Abies spp.                          fir species                     |
ACNE  Acer negundo                        boxelder, ash-leaved maple      |
```

```
00000000011111111112222222222333333333344444444445555555555666666666677777
12345678901234567890123456789012345678901234567890123456789012345678901234
```

**Figure 1. Listing of the first 12 lines of SPP.LST showing the header line, and the end-of-line bars.** The number lines in the second box depict the character column positions. Columns 1 to 4 contain the species abbreviation, columns 7–42 contain the species name, and columns 44–73 contain one or more common names.

```
ListWid = 73                                                             |
             DendroScan Version 4.5 HELP FILES                           |
                                                                         |
                                                                         |
@MENU - FILE                                                             |
                                                                         |
TIF File Convert                                                         |
                                                                         |
 This option is used to convert the 256 color grey scale files           |
 to a more compact format.                                               |
                                                                         |
@Open ISC File                                                           |
                                                                         |
      - This option is used to open ISC Files                            |
                                                                         |
@Join ISC Files                                                          |
                                                                         |
      - This option is used to append one ISC file to another.           |
```

```
0000000001111111111222222222233333333334444444444555555555566666666667777
1234567890123456789012345678901234567890123456789012345678901234567890123
```

**Figure 2. Example of a help file structure showing the header line and the end-of-line bars.** Note that help section headers all start in column 1. All other information lines contain a blank in the first column.

- The first line must contain the statement: ListWid = #. Do not insert any new lines above this statement.
- **DO NOT USE the Tab key**. This will render the file unreadable to DendroScan.
- Do not write information beyond the column specified by ListWid.

When you have completed adding new species to the file, run the program FIXLST using the following command:

**FIXLST** file name

FIXLST simply goes through the entire file and ensures that each line is the same length. If lines exceed the length specified by ListWid, they are truncated. All lines shorter than this length are space-filled until they are the correct length. The vertical bar is reinserted at the end of each line. When the program has successfully run, it will echo the following statement:

**FIX was successful**

## Species Lists

There are two species lists included with DendroScan. One, SPECIES.LST, contains all of the abbreviations of several worldwide tree species. The other, SPP.LST, is the resource file used by the DendroScan program, and in the distribution copy contains a reduced number of species (primarily North American and European species). If you wish to create a new resource list, edit the larger file and then copy it to SPP.LST.

Species lists have other properties to consider: they are arranged in a columnar format, and the lines are arranged in alphabetical order according to the species abbreviation. The following constraints, therefore, also apply when editing these files.

- Insert any new data so that the four-digit species codes remain in alphabetical order in the file.
- Do not adjust the column positions.
- Do not forget to run FIXLST after editing the file.

## Help Files

Help files follow the same convention as list files, but are written in a more-open format. Two types of data lines exist: header lines and information lines. Header lines are always left justified in the file, and contain the key words that DendroScan uses for rapid access to different help topics. Typically they start with an @ symbol.

The user might wish to edit this file to convert it to a different language. This is possible to a degree. The header lines can not be changed, as they are related to coded key words in the program itself. The help information within each section is, however, completely editable.

- Do not change the header lines. Changing them will disable some aspects of DendroScan's help facility.
- Don't forget to run FIXLST after editing the file.

# APPENDIX 7
# BASIC DENSITY CORRECTION FACTORS

## Table 1. Density correction factors for several Canadian tree species[a]

| ITRDB species code | Species name | Common name | Basic density (g/cc)[b] | Oven-dry density (g/cc)[c] | Basic density correction factor[d] |
|---|---|---|---|---|---|
| ABSP[e] | *Abies* spp. | Fir species | –[f] | – | 0.894 |
| ABAM | *Abies amabilis* | Amabilis fir | 0.360 | 0.412 | 0.895 |
| ABBA | *Abies balsamea* | Balsam fir | 0.335 | 0.367 | 0.874 |
| ABLA | *Abies lasiocarpa* | Subalpine fir | 0.331 | 0.369 | 0.913 |
| ACSP[e] | *Acer* spp. | Maple species | – | – | 0.869 |
| ACMA | *Acer macrophyllum* | Broadleaf maple | 0.466 | 0.530 | 0.879 |
| ACNE | *Acer negundo* | Manitoba maple | 0.416 | 0.501 | 0.830 |
| ACRU | *Acer rubrum* | Red maple | 0.516 | 0.586 | 0.881 |
| ACSA | *Acer saccarinum* | Silver maple | 0.461 | 0.509 | 0.906 |
| ACSH | *Acer saccharum* | Sugar maple | 0.597 | 0.702 | 0.850 |
| ALSP[e] | *Alnus* spp. | Alder species | – | – | 0.884 |
| ALRU | *Alnus rubra* | Red alder | 0.373 | 0.422 | 0.884 |
| BESP[e] | *Betula* spp. | Birch species | – | – | 0.861 |
| BEAL | *Betula alleghaniensis* | Yellow birch | 0.559 | 0.649 | 0.861 |
| BEPA | *Betula papyrifera* | White birch | 0.506 | 0.588 | 0.861 |
| CYSP[e] | *Carya* spp. | Hickory species | – | – | 0.828 |
| CYCO | *Carya cordiformis* | Bitternut hickory | 0.628 | 0.760 | 0.826 |
| CYOV | *Carya ovata* | Shagbark hickory | 0.654 | 0.788 | 0.830 |
| CADN | *Castanea dentata* | Chesnut | 0.423 | 0.466 | 0.908 |
| CHNO | *Chamaecyparis nootkatensis* | Yellow cypress | 0.419 | 0.462 | 0.907 |
| FAGR | *Fagus grandifolia* | Beech | 0.590 | 0.705 | 0.837 |
| FRSP[e] | *Fraxinus* spp. | Ash species | – | – | 0.873 |
| FRAM | *Fraxinus americana* | White ash | 0.570 | 0.650 | 0.877 |
| FRNI | *Fraxinus nigra* | Black ash | 0.468 | 0.539 | 0.868 |
| FRPE | *Fraxinus pennsylvanica* | Green ash | 0.486 | 0.556 | 0.874 |
| JGSP[e] | *Juglans* spp. | Walnut species | – | – | 0.897 |
| JGCI | *Juglans cinerea* | Butternut | 0.368 | 0.399 | 0.922 |
| JGNI | *Juglans nigra* | Black walnut | 0.546 | 0.626 | 0.872 |
| LASP[e] | *Larix* spp. | Larch species | – | – | 0.875 |
| LALA | *Larix laricina* | Tamarack | 0.485 | 0.544 | 0.892 |
| LAOC | *Larix occidentalis* | Western larch | 0.549 | 0.640 | 0.858 |
| PCSP[e] | *Picea* spp. | Spruce species | – | – | 0.894 |
| PCEN | *Picea engelmannii* | Engelmann spruce | 0.375 | 0.425 | 0.882 |
| PCGL | *Picea glauca* | White spruce | 0.354 | 0.393 | 0.901 |

**Table 1. concluded**

| ITRDB species code | Species name | Common name | Basic density (g/cc)[b] | Oven-dry density (g/cc)[c] | Basic density correction factor[d] |
|---|---|---|---|---|---|
| PCMA | *Picea mariana* | Black spruce | 0.406 | 0.445 | 0.912 |
| PCRU | *Picea rubens* | Red spruce | 0.380 | 0.425 | 0.894 |
| PCSI | *Picea sitchensis* | Sitka spruce | 0.347 | 0.394 | 0.881 |
| PISP[e] | *Pinus* spp. | Pine species | – | – | 0.914 |
| PIBN | *Pinus banksiana* | Jack pine | 0.421 | 0.454 | 0.927 |
| PICO | *Pinus contorta* | Lodgepole pine | 0.403 | 0.455 | 0.886 |
| PIMC | *Pinus monticola* | Western white pine | 0.355 | 0.398 | 0.892 |
| PIPO | *Pinus ponderosa* | Ponderosa pine | 0.438 | 0.489 | 0.896 |
| PIRE | *Pinus resinosa* | Red pine | 0.392 | 0.419 | 0.936 |
| PIST | *Pinus strobus* | Eastern white pine | 0.364 | 0.384 | 0.948 |
| PPSP[e] | *Populus* spp. | Poplar species | – | – | 0.895 |
| PPBA | *Populus balsamifera* | Balsam poplar | 0.372 | 0.416 | 0.894 |
| PPDE | *Populus deltoides* | Eastern cottonwood | 0.352 | 0.386 | 0.912 |
| PPGR | *Populus grandidentata* | Largetooth aspen | 0.390 | 0.431 | 0.905 |
| PPTR | *Populus tremuloides* | Trembling aspen | 0.374 | 0.424 | 0.882 |
| PPTC | *Populus trichocarpa* | Black cottonwood | 0.295 | 0.334 | 0.883 |
| PNSE | *Prunus serotina* | Black cherry | 0.510 | 0.623 | 0.819 |
| PSME | *Pseudotsuga menziesii* | Douglas-fir | 0.450 | 0.510 | 0.882 |
| QUSP[e] | *Quercus* spp. | Oak species | – | – | 0.875 |
| QUAL | *Quercus alba* | White oak | 0.654 | 0.775 | 0.844 |
| QUMA | *Quercus macrocarpa* | Bur oak | 0.599 | 0.674 | 0.889 |
| QURU | *Quercus rubra* | Red oak | 0.581 | 0.655 | 0.887 |
| QUVE | *Quercus velutina* | Black oak | 0.597 | 0.677 | 0.882 |
| THSP[e] | *Thuja* spp. | Cedar species | – | – | 0.947 |
| THOC | *Thuja occidentalis* | Eastern White cedar | 0.299 | 0.308 | 0.971 |
| THPL | *Thuja plicata* | Western red cedar | 0.312 | 0.338 | 0.923 |
| TISP[e] | *Tilia* spp. | Tilia species | – | – | 0.851 |
| TIAM | *Tilia americana* | Basswood | 0.360 | 0.423 | 0.851 |
| TSSP[e] | *Tsuga* spp. | Hemlock species | – | | 0.887 |
| TSCA | *Tsuga canadensis* | Eastern hemlock | 0.404 | 0.447 | 0.904 |
| TSHE | *Tsuga heterophylla* | Western hemlock | 0.409 | 0.470 | 0.870 |
| ULSP[e] | *Ulmus* spp. | Elm species | – | – | 0.851 |
| ULAM | *Ulmus americana* | White elm | 0.524 | 0.617 | 0.849 |
| ULRU | *Ulmus rubra* | Slippery elm | 0.547 | 0.643 | 0.851 |
| ULTH | *Ulmus thomasii* | Rock elm | 0.625 | 0.732 | 0.854 |

[a] Density values are derived from: Jessome, A.P. 1977. Strength and related properties of woods grown in Canada. Supply Serv. Can., For. Prod. Lab., Ottawa, Ontario. For. Tech. Rep. 21.

[b] Basic density is defined as oven-dry mass divided by green volume.

[c] Oven-dry density is defined as oven-dry mass divided by oven-dry volume.

[d] Basic density correction factor converts oven-dry density to basic density.

[e] Correction factors defined for miscellaneous species within a genus are averages of those for the particular species listed within the genus in Jessome (1977) (*see* Footnote a).

[f] Not applicable.

# INDEX

## D



## E

## J



## L



## M



## N



## O



## P

## Q



## R



## S

## T

## V



## W



## X